Winfried Schädler

Slope movements of the earthflow type –
engineering-geological investigation,
geotechnical assessment and modelling of the source areas
on the basis of case studies from the Alps and Apennines

Logos Verlag Berlin

λογος

Bibliographic information published by Die Deutsche Bibliothek

Die Deutsche Bibliothek lists this publication in the Deutsche Nationalbibliografie;
detailed bibliographic data is available in the Internet at http://dnb.ddb.de.

ISBN 978-3-8325-2593-4

Logos Verlag Berlin GmbH
Comeniushof, Gubener Str. 47,
10243 Berlin
Tel.:   +49 (0)30 / 42 85 10 90
Fax: +49 (0)30 / 42 85 10 92
http://www.logos-verlag.de

Slope movements of the earthflow type –

engineering-geological investigation, geotechnical assessment and modelling of the source areas

on the basis of case studies from the Alps and Apennines

Hangbewegungen vom Typ eines Schuttstroms –

ingenieurgeologische Untersuchung, geotechnische Beurteilung und Modellierung der Liefergebiete

anhand von Fallstudien aus den Alpen und dem Apennin

Der Naturwissenschaftlichen Fakultät

der Friedrich-Alexander-Universität Erlangen-Nürnberg

zur

Erlangung des Doktorgrades Dr. rer. nat.

vorgelegt von

Winfried Schädler

aus Gunzenhausen

Als Dissertation genehmigt
von der Naturwissenschaftlichen Fakultät
der Friedrich-Alexander-Universität Erlangen-Nürnberg

Tag der mündlichen Prüfung: 08. Juli 2010

Vorsitzender der Promotionskommission: Prof. Dr. Eberhard Bänsch

Erstberichterstatter: Prof. Dr. Michael Moser

Zweitberichterstatter: Prof. Dr. Tom Schanz
                              (Ruhr-Universität Bochum)

Drittberichterstatter: Prof. Alessandro Corsini
                              (Università degli Studi di Modena e Reggio Emilia)

Gewidmet in Dankbarkeit meinen Eltern

Foreword

First of all I would like to thank my supervisor, Prof. Michael Moser, for giving me the opportunity to work on this interesting engineering-geological research topic and the excellent mentoring at all stages of the work.

My special thanks go to Prof. Tom Schanz (Ruhr-Universität Bochum, previously at Bauhaus-Universität Weimar) for supervising this thesis in the scientific fields of Geotechnics and Numerical Modelling. It was his comprehensive support that allowed focusing the work on important interdisciplinary aspects at the interface between Geology and Engineering.

This thesis got its final character due to the support and advice of Prof. Alessandro Corsini (Università degli Studi di Modena e Reggio Emilia, Italy) who provided extensive data and experience from current and socio-economically relevant case studies.

This work is one of the results of an interdisciplinary cooperation that has been cultivated in different forms and for several years between the Engineering Geology Division of the Chair of Applied Geology at Friedrich-Alexander Universität Erlangen-Nürnberg and the Chair of Soil Mechanics at Ruhr-Universität Bochum (previously Bauhaus-Universität Weimar).

In this context I have to thank Prof. Tom Schanz for giving me the opportunity to work for four years with his research group at the Chair of Soil Mechanics at Bauhaus-Universität Weimar, enabling an interesting and fruitful exchange of information and know-how, especially with regard to the soil-mechanical aspects of the research topic and the application of Computational Geomechanics and Inverse Modelling methods. Here I have to thank particularly Dr. Jörg Meier for an intensive cooperation.

During this doctoral project, a cooperation between the above-named institutions and the Department of Earth Sciences at Università degli Studi di Modena e Reggio Emilia, Italy, was started and intensified by visits and meetings of the different research groups. This cooperation allowed for basing the work, especially the numerical modelling, on practical field and monitoring experiences and long-term and comprehensive data regarding the slope movements under investigation.

Special thanks go to the leader of the Applied Geology research group at Modena University, Prof. Alessandro Corsini, and his co-workers, Dr. Francesco Ronchetti and Dr. Lisa Borgatti, for the excellent cooperation and their advice, for example concerning the geology, hydrogeology and geomorphological evolution of the studied slopes.

Furthermore, I would like to thank Tina Knabe (previously working at the Chair of Soil Mechanics at Bauhaus-Universität Weimar) for complementing this work with a Diploma thesis on a related topic in civil engineering, namely the implementation of active structural mitigation measures into one of the geotechnical/numerical models that were developed in this doctoral thesis and the evaluation of the effects of these constructions.

Thanks also to Andreas Ressle and Harald Meier (Engineering Geology Division at Erlangen University), for their support during the fieldwork in the Lahnenwiesgraben testing area in the Bavarian Alps.

Soil-mechanical laboratory tests and classifications were performed at Bauhaus-Universität Weimar, at the Laboratory of Soil Mechanics and at the geotechnical laboratory of the MFPA Material Research and Testing Institute. I would like to express my gratitude to the staff of these institutions for their help and technical support during laboratory work.
For the Laboratory of Soil Mechanics, special thanks go to Prof. Tom Schanz, Gabriele Tscheschlok and Frank Hoppe.
As representatives for the team of the MFPA, Prof Karl-Josef Witt and Klaus Lemke are thanked.

And of course I want to thank all my colleagues and friends in Weimar, Modena and Erlangen for their help and encouraging remarks during the whole time of my work.
Further thanks go to Geeralt van den Ham and Prof. Joachim Rohn for giving some hints and tips from the standpoint of their own doctoral theses about earthflows.

The following public authorities and their representatives are thanked for having supported my work:
The Geological Survey of the Autonomous Province of South Tyrol (Landesamt für Geologie und Baustoffprüfung) at Kardaun, Italy, particularly Dr. Ludwig Nössing and Dr. Volkmar Mair. The Forestry Office at Farchant, Landkreis Garmisch-Partenkirchen, Germany, particularly Claus Walcher. The Research Institute for Hydrological and Geological Hazard Prevention IRPI of the National Research Council CNR, Padova, Italy, particularly Dr. Sandro Silvano and Dr. Gianluca Marcato.

I also would like to thank the Konrad Adenauer Foundation, Dr. Daniela Tandecki of the Deutsche Graduiertenförderung is being named here as a representative, for granting me a postgraduate scholarship, which allowed me to focus in detail on my scientific project. I therefore feel very thankful about the confidence in me and the interest for the topic of my work.
The German Academic Exchange Service (DAAD) and the Association of the Rectors of the Italian Universities (CRUI) are acknowledged for funding travelling expenses for me and my colleagues through a VIGONI exchange project.

With all my heart I thank my parents and my parents-in-law for supporting my work in many ways, and especially my wife, Verena Schädler whom I met thanks to this thesis.

Contents

1. Introduction

1.1. Description of the topic and the motivation for this work

Landslide risk is a problem in many mountain areas. In some regions of the world, landslides of the earthflow-type represent an important and ongoing process of landscaping (e.g. *Mackey et al. 2009*). Repeated reactivations and acceleration phases of these movements have been recorded and dated throughout the Lateglacial and the Holocene (i.e. during the last 15000 years) and are observed also in present times (e.g. *Bertolini and Pellegrini 2001*).

While in some countries – for example in Germany – very small areas are concerned, landslides of the earthflow-type are very common in Italy, particularly in the Dolomites and in the Northern Apennines: In the Emilia Romagna Region, they represent about 80% of all known landslides (e.g. *Ronchetti et al. 2007*). Earthflows are among the largest landslides worldwide, e.g. the Slumgullion earthflow in Colorado, United States of America, with a length of seven kilometres (e.g. *Savage et al. 1996*). Also in the Alps and Apennines, they are up to several tens of metres deep and several square kilometres large (e.g. the Stambach earthflow in Upper Austria, *Rohn 1991*)

In the affected regions, these landslides represent a phenomenon of major economic relevance, as they cause continuous damaging of infrastructures and in several cases pose a potential threat to human settlements. The active stage of an earthflow can lead to a total destruction of all buildings and of the entire infrastructure situated on the moving mass (e.g. *Raetzo and Lateltin 1996*). In such critical situations, the largest potential dangers, not only in terms of economic losses but also for human lives, arise not from the earthflow itself: Frequently, devastating debris-flows and soil slips may develop when the earthflow gets in contact with larger amounts of water and where the front of an advancing part of the mass gets oversteepened as a consequence of the movements and deformations. Accelerated earthflow movements can dam up a river and seriously endanger the settlements downstream, since a possible failure of such a natural dam can cause severe floods, debris-flows or mud flows (e.g. *Raetzo and Lateltin 1996, Vulliet and Bonnard 1996*).

Generally, the mitigation of landslide risk can be based on the reduction of vulnerability or on the stabilisation of the hazardous phenomena. The first approach pivots around the enforcement of rules for land use and construction, the set-up of monitoring and early warning systems and, as ultimate solution, the relocation of the elements at risk. The second answer to the problem entails countermeasures on the slopes that can eliminate or reduce the physical causes and the effects of the instability processes. More or less invasive works on slopes are often inevitable when roads, buildings, services and finally people are already located inside or nearby the hazard zones. When considering the landslide type studied in this work, the latter is quite often the case, because for geological reasons, these landslides develop in the more gently-sloped parts of mountainous regions (e.g. *Bertolini et al.*

2005), which sometimes have been chosen as favourable location for villages or are crossed by important roads.

Deep-scated earth-slides/flows are commonly mitigated by means of deep drainage systems, e.g. sub-horizontal drains, large-diameter wells, micro- and macro-drainage tunnels, which are often combined with bored pile retaining walls (see e.g. *Holtz and Schuster 1996*). As such interventions usually are expensive and have a considerable environmental impact, a dependable pre-evaluation of the benefits obtained by these measures is of great interest. In this context, physically-based numerical simulations of the landslide behaviour that are carried out at the slope-scale represent an important tool: If such numerical models are sufficiently realistic, they allow for integrating different combinations of countermeasures into the simulation for pre-evaluation of their possible effects, and compared to the eventually planned real building measures, these simulations are cheap.

1.2. Definition and classification of the slope movements under investigation

In the literature of natural sciences and engineering, slope movements, commonly and in a general sense also referred to as "landslides" are not classified and termed identically by all authors. Different classifications were published (e.g. *Varnes 1978, Cruden and Varnes 1996, Hutchinson 1988, Hungr et al. 2001, Moser and Üblagger 1984*). The usage of certain terms differs in different languages and countries. In this context, it has to be considered that each landslide is a unique phenomenon, never identical to any other. Furthermore, different aspects of research lead to the setting up of different terminologies. Thus, in order to avoid misunderstandings or confusion with other phenomena, a definition of the slope movements under investigation in this work is made at this point:

Definition:
This thesis deals with natural slope movements, which are of continuous and long-lasting nature, i.e. taking place over years to thousands of years. These movements affect and are due to the presence and formation of thick soil covers, i.e. metres to tens of metres thick, covering in-situ rock masses. These soil covers are formed mainly by the weathering of the nearby or underlying rock-masses and consist of coarse components, i.e. coarse gravel, cobbles and boulders, in a fine-grained clay-rich matrix. The overall phenomenon is characterised by pronounced stages of activity that are marked by the varying speed of the movements, which are relatively slow compared to other landslide types, i.e. ranging from mm/year to some tens of meters per day.

In literature, these phenomena are mostly found under the term of "earthflows" (also: "earth-flow" or earth flow). However, the term "earthflow" as well as the German word "Schuttstrom" is sometimes applied to much faster phenomena than the ones studied here:

For example, *Hanisch 2004,* applies the term "Schuttstrom" for debris carried in suspension, furthermore, as remarked by *Comegna 2005,* when following the classification of *Varnes 1978*, also the landslides that are observed in the Quick-clays of Norway are termed "rapid earthflows", although they are completely different from the ones studied here, with respect to the formation and related properties of the soils concerned, i.e. they involve clays that were sedimented very recently (i.e. only some thousands of years ago) in a marine environment. Instead of "earthflow" many authors prefer the term "mudslide" (e.g. *Hutchinson 1988*) stating that movements chiefly take place by sliding on discrete shear-zones. For the same reason, the studied landslides are often referred to as "earth-slides".

Generally, it has to be considered that slope movements of the type investigated in this work take place over long time spans. Movement type and velocity vary with time. Thereby remobilisations, i.e. accelerations, can be partial or total (e.g. *Bertolini and Pellegrini 2001*, *Ronchetti et al. 2009 A*). This means that, as an example, when applying the internationally most widely used classification for landslides of *Cruden and Varnes 1996* to the phenomenon studied here, the same material at the same position on an affected slope may move as an "earth slide", showing "very slow" translational, partly rotational displacements. A few months later, it may partly be mobilised as a "slow" "earth flow" and some additional months later, all these movements may come to a nearly complete stop, except for "extremely slow" movements without a shear-zone developed, i.e. this would again be an "earth flow", and so on. Instead of the time specifications "months", also "years" or "thousands of years" may be used, in any order, in the described example. Thus, as shown by the example, a combination of various different terms is needed for describing the landslide only at a certain position within the landslide area. Furthermore, in many cases, in the upper part of the affected slopes, a transition process from a disintegrating soft rock mass to a soil (in engineering-geological terms) is observed. Following the classification of *Cruden and Varnes 1996,* these movements are to be classified partly as "rock slides". For this reason, *Picarelli et al. 2005,* although recommending this classification as a "rational and well-organized process" remarked that it "results in long names".

As a consequence, in this thesis, when shortly referring to the studied phenomenon as a whole, the term "earthflow" is used (quasi describing a worst-case scenario for the affected slope area), knowing that flow-type movements take place only during certain times and in certain parts of the area under investigation. In contrary, when a limited area at a certain time is considered, the terminology of *Cruden and Varnes 1996* is used, with the implicit understanding that only phenomena covered by the definition given in this Section (on page 2) are meant. Where this classification results in too long names or where the term "earthflow" may cause confusion, the very general and neutral term "landslide" is preferred.

1.3. Main thematic focus of the work

The main thematic focus of this work is on the investigation of the so-called **source areas** located in the upper parts of the affected slopes. For the future development of this type of slope movements, the source areas are of particular importance: Firstly, from these areas the landslide is provided with new material. Secondly, several important processes which are able to cause reactivations or accelerations of the slope movements (see Chapter 2, pages 49 to 52 for details) most often start from the source areas. By means of two case studies, one from the Apennines and one from the Alps, selected aspects of the source areas are investigated in this thesis.

As **first case study**, the source area of the frequently reactivated Valoria earthflow is chosen to study the medium-term history (years to decades) of the slope morphology in the crown and head zone by means of a qualitative reconstruction in a numerical simulation. Using this tool, different phases of activity and critical events are analysed together with field- and monitoring-data in order to obtain an improved geological, kinematical and mechanical understanding of the slope evolution.

In the **second case study**, an important part of the source area of the Corvara earthflow is selected to analyse seasonal changes of monitored movement rates during a relatively calm phase of the landslide's history together with on-site meteorological data. A numerical model based on simplifying geotechnical and hydrogeological assumptions is set up. Inverse modelling methods are used to calibrate this numerical model and to back-calculate important material parameters on the basis of field-measurements. Active structural mitigation measures are then implemented into the calibrated numerical model in order to evaluate whether such measures are likely to stabilise the studied source area, i.e. to stop the observed slope movements approximately or not.

1.4. State of the art and literature review

1.4.1. Earthflows in the literature of Geosciences since 1960

Geological characteristics of earthflows were studied by a large number of authors. Detailed studies were carried out especially in North America, e.g. *Bovis 1985, Bovis 1986, Fleming et al. 1988 A* or *Keefer and Johnson 1983*, and in Italy, e.g. *Giusti et al. 1996* or *Comegna et al. 2007*, in the Southern Apennines, e.g. *Dall'Olio et al. 1988, Panizza et al. 1996 and 1997 or Soldati et al. 2004*, in the Alps, e.g. *Betelli and De Nardo 2001* or *Bertolini et al. 2005*, in the Northern Apennines. As regards German-language publications that studied earthflows in Austria, Northern Italy and Switzerland from a geological viewpoint, *Fischer 1967, Baumgartner and Mostler 1978, Bunza 1978, Baumgartner 1981, Bammer 1984, Rohn 1991, Raetzo and Lateltin 1996* or *Raetzo et al. 2000,* can be mentioned, as examples.

With respect to landslides of the earthflow-type, **Geomorphology, history of landslide activity** and the **role of climatic changes** in this context were investigated e.g. by *Bovis 1985, Panizza et al. 1996 and 1997, Corsini et al. 1999, 2000, and 2001, Bertolini and Pellegrini 2001, Borgatti 2004, Soldati et al. 2004, Borgatti et al. 2007 A,* or by *Giusti et al. 1996* (mainly geomorphologic features) or *Raetzo and Lateltin 1996* (only history of activity).

Monitoring data on movements and hydrogeological conditions linked to such landslides have been published and interpreted e.g. by: *Keefer and Johnson 1983, Jedlitschka 1984, Bovis 1985 and 1986, Conte et al. 1991, Rohn 1991, Berti et al. 1994, Froldi and Lunardi 1994, Bonnard et al. 1995, Angeli et al. 1996 A and 1998, Giusti et al. 1996, Savage et al. 1996, Wasowski 1998, D'Elia et al. 2000, Pellegrino et al. 2000, Raetzo et al. 2000, Corsini et al. 2005, Ronchetti et al. 2006, van den Ham 2006, van den Ham et al. 2009, Borgatti et al. 2007 B, Comegna 2005, Ronchetti 2007,* or *Keller 2009*. Monitoring data on the movement of earthflows were, for example, also presented (among data from other landslide types), in detailed reports of the University of Lausanne Switzerland, e.g. *VERSINCLIM 1998* and *D.U.T.I. 1983* (various authors). Generally, for measuring surface displacements, photogrammetric methods, wire-extensometers or steel measuring tapes have often been used, while in more recent times, theodolites with electronic distance meters (EDM) and GPS (Global Positioning System)-measurements have added to the group of techniques applied. Many of the above-mentioned authors used different techniques simultaneously in order to be able to check the results or to compensate the disadvantages of one method by the advantages of another one. This has turned out as good practice, because due to the large and often unforeseeable deformations of the ground, malfunction or destruction of measurement devices or benchmarks can easily occur. A promising new method for measuring surface displacements is ground-based SAR-interferometry (Synthetic Aperture Radar): Its application was demonstrated by *Tarchi et al. 2003*. The measurement method allowed for obtaining multi-temporal surface deformation maps of the entire source area of the Tessina earthflow. Thereby displacements in the mm-range were detected at a high resolution. Sub-surface displacements were nearly exclusively measured by means of inclinometers. Recently, also TDR-cables (Time Domain Reflectometry) have been used for the detection of shear-zones in earthflow deposits, e.g. by *Corsini et al. 2005*.

Data on **geotechnical and soil mechanical properties of the materials** involved in slope movements of the earthflow-type have been reported predominantly from Italy, e.g. by *Angeli et al. 1991, 1996 A , B and 1998, Bertini et al. 1991, Cancelli et al. 1991, Berti et al. 1994, Conti and Tosatti 1994, Froldi and Lunardi 1994, Guadagno et al. 1994, Giusti et al. 1996, Bertolini and Gorgoni 2001* or *Comegna 2005,* but also from other regions of the world, e.g. by *Hutchinson and Bhandari 1971* (Great Britain), *Keefer and Johnson 1983* (United States of America, California), *VanDine 1983* (Canada, British Columbia), *Jedlitschka 1984* (Austria), *Bovis 1985 and 1986* (Canada, British Columbia), *Fleming et* *al. 1988 B* (United States of America, Utah), *Rohn 1991* (Austria) and *Rohn et al. 2005* (Austria), *Nozaki et al. 1994* (Japan), *van den Ham 2006* (Austria) or by *van Asch et al. 2007 B* (France). Comprehensive studies with respect to Geology, monitored slope movements and soil mechanical properties of the materials were carried out on several landslides of the Emilia Apennines (mostly of the earthflow-type), which resumed activity in the 1994-1999 period (*Bertolini, Pellegrini and Tosatti – editors – 2001, collection of contributions of numerous Italian authors*).

Reflections about the **mechanisms of earthflow-type landslides**, i.e. the mechanical causes and characteristics of the deformations observed in the soils and soft rocks involved, were published e.g. by *Haefeli 1967, Hutchinson and Bhandari 1971, Scheidegger 1975, Fleming et al. 1988 B, Leroueil et al. 1996, Picarelli et al. 1995, Comegna 2005* and *Comegna et al. 2007, van Asch et al. 2007 B*, or *Hofmann 2009*.

1.4.2. Numerical modelling of landslides in general and of earthflows in particular

At the present state of the art, understanding, forecasting and controlling the hazard associated to the different types of landslides is still a largely empirical task, integrating both qualitative and quantitative analyses of datasets pertaining to several disciplines (e.g. Geology, Geomorphology, Hydrogeology, Hydrology, Geophysics, Geotechnics etc.) (see e.g. *van Asch et al. 2007 A*). These analyses can be performed on several spatial and temporal scales, mainly depending on the objectives of hazard assessment (e.g. *Aleotti and Chowdhury 1999, Crozier and Glade 2005,* and *van Westen et al. 2006*). Some answers to the important questions (how, where and when?) can be given by the so-called engineering-geological judgment. But in many instances, there is a need for numerical modelling, because purely qualitative or descriptive answers are often inadequate and/or can not capture the variability of different influence factors involved in slope stability problems (see e.g. *Bromhead 1996, Corominas and Ledesma 2002,* or *van Westen et al. 2006*). Therefore, since the 1970s, when the pioneering studies on the numerical modelling of landslides were published (e.g. *Kalkani and Piteau 1976,* on a toppling phenomenon, *Krahn and Morgenstern 1976,* on a large rock slide, or *Radbruch-Hall et al. 1976,* on a Sackung phenomenon), great efforts have been put into this subject.

Numerical modelling methods and tools that have been applied to landslides are very different, ranging from statistical techniques, such as multivariate analysis, which is generally used to assess the susceptibility of slopes to landslides at a regional scale, to process-based approaches, which are applied also on a local scale, e.g. limit-equilibrium methods, continuum and discontinuum methods or hybrid continuum/discontinuum methods. A quite large number of research and commercial computer codes allow for the simulation of slope stability and landslide phenomena in different conditions with

6

respect to space (2D or 3D) and time (constant time or time-dependent), and based on different mechanical approaches (see e.g. *Bromhead 1996* and *Stead et al. 2006*). As an example, various numerical techniques, namely continuum, discontinuum and hybrid methods, were applied by *Eberhardt et al. 2004* to the case study of the Randa rock slide (Switzerland), to demonstrate the evolution of failure in massive natural rock slopes as a function of slide plane development and internal strength degradation. Research continues on the development of new numerical modelling methods and on the extension of the capabilities of the existing methods. For example, *Crosta et al. 2003* presented a new 2D/3D- Finite-Element code and demonstrated that it is able to analyse both slope stability or relatively small deformations inside a rock-slope and the run-out of a failure mass, characterised by very large displacements, using very different material laws – all this in one and the same model – and suggested the application of the presented methods to rock slides, rock avalanches and debris avalanches.

Numerical models describe the relationships between the predisposing and triggering factors, as model input and boundary conditions, whereas the responses of the slopes are presented as model outputs. However, in most cases, building up an effective numerical model is difficult, due to the complex nature of geologic materials and the four-dimensional (space and time) pattern of slope movements (e.g. *Brunsden 1999*). The exploitation of geomorphological evidence and supporting monitoring data is therefore a key issue in all the phases of modelling, both as input and as feedback (e.g. *Corominas and Ledesma 2002*).

Numerical modelling has been applied **to earthflow-type landslides** by various authors and using different approaches. Several examples are given in the following: *Berti et al. 1994* applied a Finite-Differences model using an elastic - ideally plastic material model with the Mohr-Coulomb failure criterion to an earthflow-type landslide at Acquabona (Northern Apennines, Italy). On the basis of their geotechnical model, these authors calculated the Geometry of the main shear-zone and the related distribution of displacements, which were compared to field measurements. The piezometric levels measured in a large number of boreholes before and after the realisation of mitigation measures were used as input in stability and deformation analyses to check the effectiveness of the mitigation measures carried out. *Angeli et al. 1996 A and B*, also *Angeli et al. 1998,* combined a hydrological model with a visco-plastic stability model to simulate the velocity of landslide movement along a secondary shear zone of the Alverà earthflow (Italian Alps). A similar viscosity model was applied by *van Asch et al. 2007 B* to the La Valette earthflow (French Alps). *Vulliet and Bonnard 1996* applied a three-dimensional viscous numerical flow model describing the slow motion of natural slopes which may creep and slide along their basal surface. For simulating the properties of the moving material, a single-phase model considering an incompressible non-Newtonian viscous body was used. Calculations were performed by means of the Finite-Differences method. This model was used for predicting the shape and height of the dam formed by the advancing of the Falli Hölli earthflow

(Switzerland) towards the Höllbach-river in 1994. *Avolio et al. 2000* adapted a Cellular Automata model, which was originally developed for much faster phenomena, such as debris flows, to the Tessina earthflow (Italian Alps). Calibration of the model was obtained from the simulation of an event that had occurred in 1992. The calibrated model was then used for forecasting the trajectory and the extensions of the earthflow during possible future reactivation scenarios.

Picarelli et al. 1995 used Finite Elements to simulate the excess pore pressures induced by an undrained thrust in the upper part of an earthflow body and the displacement profiles resulting from this pressure increase in different parts of the slope. An elasto-plastic constitutive law with the Mohr-Coulomb failure criterion was used. Continuing the afore-mentioned work, *Comegna 2005* and *Comegna et al. 2007,* based on the vast experiences that were meanwhile gathered during long-term investigation and monitoring of some earthflows in the Basento Valley and in the Miscano valley in Southern Italy (see already *Giusti et. al 1996*) and on comprehensive laboratory testing, built a 2D Finite-Element model of the track zone of the Masseria Marino earthflow to study the mechanisms behind the observed pore pressure oscillations and acceleration phases. The Mohr-Coulomb Model and the Soft Soil Model (*PLAXIS Material Models Manual*) were used as constitutive models. Pore pressure fluctuations due to climatic conditions, as well as excess pore pressures due to a sudden load acting onto a part of the earthflow body (only in *Comegna 2005*), due to a distributed load moving along the earthflow body (only in *Comegna 2005*), and the excess pore pressures following the development of a new crack through the earthflow body were simulated and compared to time series of piezometer measurements in different parts of this earthflow body. The calculated deformations related to these changes of the pore pressure distribution were analysed together with the monitored distribution of displacements in the field. Summing up, the work of *Comegna 2005* and *Comegna et al. 2007* is actually the numerical study on earthflows, which is based on the most detailed field, laboratory and measurement data, that could be found in literature.

Keller et al. 2005 carried out disposition modelling using the statistical Certainty-Factor Method for a catchment in the German Alps, which was mapped in the field at a scale of 1:5000, with respect to bedrock geology, hydrology and type and thickness of the soil covers. Thereby one part of the catchment was used as training area for the model to predict the areas susceptible to shallow earthflow phenomena in the neighbouring test area. *Van den Ham 2006* simulated inclinometric deformations measured at the Stambach earthflow (Upper Austria) by means of a continuous "solid" visco-hypoplastic body on a rigid bottom. The Finite- Element method was used and results were discussed together with the measured displacement profiles. *Borgatti et al. 2007 C,* as well as *Marcato et al. 2009* used the Finite-Differences method for the simulation of slope movements in the source areas of the Corvara earthflow and the Tessina earthflow (Italian Alps), focusing on mitigation issues and on the possible future development of these landslides. The above-mentioned authors applied an elasto-visco-plastic material model to simulate the material behaviour along the sliding zones. Plasticity and viscosity parameters were determined in a back-analysis by fitting computed and measured displacements in fully saturated conditions via trial-and error procedure.

1.4.3. Numerical modelling applied for reconstruction of the slope history related to landslides

Numerical modelling has been applied by several authors to describe the long-term evolution of slopes during the Holocene. For example, *Agliardi et al. 2001* incorporated groundwater effects into a finite-difference analysis of deep-seated rock slope deformation in the Rhaetian Alps near Bormio (Italy), at its onset during the Lateglacial. Using a similar approach, *Baron et al. 2005* simulated the effects of extreme precipitation during the humid phases of the Holocene or after the melting of permafrost during the Lateglacial, investigating the possible reasons for the triggering of large-scale slope failures in the rock masses of the Magura Nappe in the Czech Republic. *Brideau et al. 2009* highlighted the importance of tectonic structures and associated rock-mass discontinuities on catastrophic slope failures, carrying out numerical modelling on the case studies of the Hope Slide (Canada, British Columbia 1965) and the Randa Rockslide (Switzerland, 1991), based on engineering-geological mapping, laboratory testing and analysis of data via GIS (Geographic Information System). *Spickermann et al. 2003* performed a Finite-Element analysis involving the determination of initial stresses resulting from the quaternary-geological history of a potentially unstable slope in the Swiss Alps, accounting also for the degradation of rock properties and the existence of disturbed zones at depth. Numerical modelling has been used by *Hermanns et al. 2006* to analyse stresses and displacements resulting from the rock-slope failures at Köfels (Austria) and Tafjord (Norway). Two-dimensional Finite-Element models of the rock-slopes prior to and following the Köfels and Tafjord failures indicate that the sudden release of large rock masses causes new zones of instability and weakness, that further reduce slope stability. On different types of landslide phenomena, some authors have used numerical simulations for retrospectively describing and back-analysing singular catastrophic events and for testing hypothetical initiation processes: For example, *Ghirotti et al. 2007* exploited high-resolution topographic data obtained by Laser-scanning to model a rock-slide event which has occurred in 2005 in the Northern Apennines of Italy.

1.4.4. Application of Inverse Modelling strategies to geotechnical problems

For many years, the application of inverse analyses for the calibration of numerical models, for the identification of parameters that can not easily be determined by laboratory experiments and to gain information on model behaviour has been a common approach in many engineering fields. For example, inverse modelling strategies are widely used in hydraulics, damage analysis and structural dynamics. For the above described tasks, Applied Mathematics provides a variety of different optimisation schemes and algorithms. In recent years, due to the availability of sufficiently fast computer hardware, inverse parameter identification strategies and optimisation procedures are more and more frequently used also in Geotechnics and Engineering Geology. Applications were described by many authors, for example, in the calibration process of geotechnical models (*Calvello and Finno 2004*), or to identify hydraulic parameters from field drainage tests (*Zhang et al. 2003*). Already *Ledesma et al. 1996 A* and *Gens et al. 1996* applied gradient methods to a synthetic and a real example of a tunnel drift simulation. Also during the excavation of a cavern in the Spanish Pyrenees the above mentioned group of authors was applying gradient methods for the identification of geotechnical parameters (*Ledesma et al. 1996 B*). *Malecot et al. 2004* used inverse parameter identification techniques for analyzing pressuremeter tests and Finite-Element simulations of excavation problems. For the identification of soil parameters, also genetic algorithms were studied (*Levasseur et al. 2007 A, Levasseur et al. 2007 B*). *Feng et al. 2006* used an inverse technique for the determination of the parameters of viscoelastic constitutive models for rocks, based on genetic programming and a particle swarm optimisation algorithm. In the field of geoenvironmental engineering, *Finsterle 2006* examined the potential use of standard optimization algorithms for the solution of aquifer remediation problems in three-phase and three-component flow and transport simulations of contamination plumes. As a different aspect of parameter identification, *Cui and Sheng 2006* determined the minimum parametric distance to the limit state of a strip foundation by optimizing a reliability index. In *2006, Schanz et al.* applied particle swarm optimization techniques to geotechnical field projects and laboratory tests, namely, a multistage excavation and the desaturation of a sand column. *Meier et al. 2008* presented an inverse parameter identification technique, based on statistical analyses and a particle swarm algorithm, to be used in the calibration process of geomechanical models. Its application was demonstrated by means of strongly simplified numerical models of two laboratory tests and of a slope section through the source area of the Corvara earthflow. Furthermore, *Meier 2008* evaluated a series of different optimisation algorithms with respect to their applicability in combination with this technique and for a wider range of geotechnical applications.

1.4.5. Numerical simulation of landslide mitigation measures

Numerical modelling of landslides is generally done to test and improve the understanding of these hazardous phenomena. One principal motivation behind this kind of research is that the obtained knowledge can be used during the planning of eventual passive or active mitigation measures. Consequently, there are also a growing number of case studies, where mitigation measures have been incorporated into the simulations of landslides or landslide-prone slopes and model behaviour was analysed in relation to these countermeasures. Regarding this topic, a variety of methods applied and strategies followed can be found in literature. Some examples are given in the following.

Firat 2009 analysed the effect of piles on a model slope susceptible to landsliding by calculating the stability of a 2D section with 9 different methods for slope stability analysis and using two different approaches for the computation of lateral forces, considering plastic deformation and viscoplastic flow. *Biavati and Simoni 2006* used a Finite-Element model to simulate the infiltration rate from the topographic surface and the transient hydraulic response at depth in unstable natural clay slopes. By this means, they analysed how these processes influence the effectiveness of drains, considering also the implications of the obtained results for slope stability. Three-dimensional Finite Element analyses for the simulation of landsliding on slopes stabilised with skewed anchors were reported by *Hazarika et al. 2008*. Numerical modelling was carried out parallel to a laboratory experiment performed on a small-scale reproduction of the slope. Thereby stability and displacements of the slope model with and without anchoring were discussed. By integrating a coupled hydrogeological stability model into a GIS system, *van Beek 2002* simulated the spatial distribution of landslide susceptibility in a catchment in south-eastern Spain. The model was calibrated by means of field data and was then used for evaluating the effect of different land-use scenarios on the distribution of predicted slope failures (a short description can be found in *van Beek and van Asch 2004*). In a case study on large debris avalanches in the Italian Prealps, the effectiveness of planned passive countermeasures was evaluated by *Crosta et al. 2006*. Slope stability analyses and a numerical run-out model were applied to identify potential instabilities and to predict their run-out patterns for possible scenarios, which were then analysed in relation to the envisaged countermeasures. *Marcato et al. 2008* used a Finite-Differences model in combination with an elasto-viscoplastic material law for the simulation of a large rotational slide in the Carnian Alps. After calibrating the model based on inclinometric measurements and GPS (Global Positioning System) data, two different types of drainage measures and a retaining wall were implemented and their effects were discussed. The finite differences method and an elastoplastic constitutive model were applied by *Borgatti et al 2008* for the modelling of a large composite landslide of the earthflow type in the Northern Apennines (the Ca' Lita landslide). A complete reconstruction of landslide development between 2002 and 2008 was attempted, including a detailed modelling of the situations before, during and after the installation of retaining walls and drainage systems, in order to assess the efficiency of the mitigation system and to propose further countermeasure works in different scenarios. *Bonzanigo et al. 2001* investigated the case of a large creeping landslide in metamorphic rock masses, located in the Swiss Alps, by means of a Distinct Element model. A coupled hydro-mechanical approach was adopted to simulate the effects of a drainage adit, which was built in 1995 and whose completion was followed by a significant decrease of slope movements. On the basis of their numerical model, these authors explored the related mechanisms and discussed the influence that drainage may have had on the stabilisation of the slope.

1.5. Contents and aims of this work

At the beginning of this work, the earthflow phenomenon is described and investigated in a general manner. Earthflows are studied with respect to their occurrence, morphological features, the properties of the materials involved, the hydrogeology of the affected slopes and the characteristics of the movements taking place. This is done based on a literature review, on detailed field work carried out by the author of this thesis in a testing area in the German Alps, and on the classification of soil samples taken of different types of earthflow-prone materials at different sites in the alpine region. The results of this part of the work are presented in **Chapter 2**.

In **Chapter 3**, the results of the reconstruction and numerical modelling of past evolution phases in the source area of the Valoria earthflow (Northern Apennines, Italy) are presented. The Valoria landslide has shown frequent reactivations and acceleration phases during the last 60 years. Since 2001, every reactivation was characterized by the retrogression of a certain part of the head and crown area, which was linked to the advancing destabilization and reshaping of these areas. This slope region, the actual upper source area, has been largely intact throughout the Holocene and started being implicated into the retrogression of the crown zone of the large prehistoric and complex earth slide/flow a few hundred years ago (*Guerra 2003, Bertolini 2003, Ronchetti 2007*). The reactivations of the head area triggered the advancing of the earth flows in the middle and lower part of the slope. In order to evaluate the masses that can be mobilised by possible future reactivations it is therefore necessary to understand the evolution of the head area.

As a consequence, starting from 2001, when the first critical event had affected this area, a large amount of valuable field and monitoring data were collected: Detailed geological and geomorphologic maps, borehole cores and refraction seismics gave information on the subsurface geometry. Laboratory data of soil samples and the classification of rock masses in outcrops allow for specifying the properties of the materials involved. Numerous measurements, obtained with different instruments and – due to the repeated destructive accelerations of the landslide – for different time spans, yielded data on surface and subsurface displacements and on the hydrological conditions. The changes in slope morphology associated with the critical events were documented precisely, always covering the situation before and after the event, by Digital Elevation Models (DEMs), maps, profiles and photographs.

Owing to the complexity of the phenomenon, the analysis of these data requires interpretation, which is neither straightforward nor simple. On the contrary, the data can be read in different ways and quite differing interpretations could be justified by them. For this reason, it is very important that the plausibility of the engineering-geological hypotheses made on the basis of the field data and of the resulting geological, kinematical and mechanical understanding of the present situation is checked extensively and carefully.

The approach followed in this work for carrying out in practice this plausibility test is that of a reconstruction of the past evolution of the landslide in a geotechnical and numerical model. In doing so, all collected information and the derived hypotheses on geometry, material properties and water pressures are included as input into the model, calculating as output deformation rates for different stages of the slope evolution. These are then compared qualitatively with the displacement rates measured by the monitoring instruments in the field. By this means, it was verified if with the current knowledge and hypotheses the following important features of the studied landslide area can be explained:

The differing velocities of the movements

      (1) at different times / in different seasons and

      (2) in different regions of the slope, also

      (3) the combination of both aspects, that is, the changes in spatial distribution of higher and
           lower movement rates with time, moreover,

      (4) the initiation of the past critical events, why they happened at a certain time, and

      (5) where the movements started accelerating to which extent, at the beginning of a
           particular event.

This type of modelling approach has several advantages: While the different phases of the slope evolution can be played through, uncertain engineering-geological assumptions can be modified. Thereby the effect of altering particular hypotheses or parameters on the entire system can be evaluated. Also the effect of modifying assumptions concerning the situation in a certain time span while maintaining the general assumptions can be studied with respect to the other simulated periods. Thus, the numerical simulation of the slope history is seen and used here as a powerful tool for evaluating how data of very different type, of differing spatial or temporal scope and of different quality, composed of measured values, well-established facts, experiences obtained from comparable sites and site-specific assumptions, all fit together to a realistic although simplified overall picture of the circumstances in the field. Modelling was performed in the framework of continuum mechanics applying the Finite-Element Method. Calculations were carried out by means of a well-established commercial code, using the elastic-viscoplastic Soft Soil Creep Model (*Vermeer and Neher 1999*) as constitutive model for the materials of the shear-zones. The basic assumptions underlying the simulations state that the evolution of the investigated slope area is determined

      (1) by changing groundwater conditions,

      (2) by the progressive reduction of friction and cohesion along the shear-zones,
          which is linked to the recent and ongoing sliding processes, and

      (3) by the related changes in slope morphology.

In order to account for different possible interpretations of the field- and monitoring-data with respect to the subsurface geometry, two different 2D geometry models were set up along a representative section through the upper source area, resulting in two different numerical models. The first model is simplified and more robust. The second one is more complex and detailed, but implies a higher number of uncertain assumptions. This strategy on the one hand provides an impression on how far the results depend on plausible but differing geometry assumptions and prevents from interpreting the results too extensively in detail. On the other hand, it helps to ensure that the more general results are not controlled by peculiarities or artefacts of one specific numerical model.

In **Chapter 4**, a concept for the analysis and numerical modelling of the hydro-mechanical behaviour of the source areas of large earthflows is presented and applied to the Corvara case study (Dolomites, Italy). The specific intention of the presented approach is to render possible the simulation and pre-estimation of the effectiveness of envisaged drainage measures and slope reinforcements. It consists of two parts: forward-calculations and inverse-modelling.

The concept for the forward-calculations has four different components:
- First of all, a geometry model that describes the positions and the course of the most important shear-zones of the studied source area and the geometry of the blocks and layers between these zones.
- Second, an adequate material model that takes into consideration weights and stiffnesses of these blocks and layers.
- Third, a creep model that describes the time-dependent material behaviour of the shear-zones.
- And finally, a hydrogeological model that describes the transient distribution of pore water pressures inside the slope with time, which is used as an important boundary condition.

These components are needed to simulate the deformations of the slope by calculating the spatial and temporal distribution of displacement rates. This can be done with different numerical methods. Here, a continuum mechanical approach with the Finite-Element method is used. Whatever the followed numerical approach, constitutive models are needed to describe the material behaviour of the soils and rocks involved and these constitutive models require sets of material parameters to be specified. In order to derive the key input parameters which give the best possible simulation of reality, the Finite-Element model is calibrated against displacement rates measured in the field, by iteratively varying the material parameters of the shear-zones, using an inverse approach. For this task, the inverse modelling concept of *Meier et al. 2008*, which was already mentioned on page 10 is applied. It consists of inverse procedures as they are used in many engineering fields, which have been adapted and tested especially for geotechnical applications. The technique is based on statistical analyses and selected optimisation algorithms. The inverse modelling strategy is further used to gain statistical information on model behaviour, on the sensitivity of model parameters and on the quality of the obtained calibration. Once the hydro-mechanical behaviour of the slope is simulated with a sufficient level of accuracy, the model is suitable for incorporating mitigation works and for evaluating the achievable effects.

On the basis of the Corvara case study, the requirements, difficulties and problems as well as the advantages and benefits of the proposed numerical modelling concept are discussed. In addition, by giving reference to a study that is based on this work (*Knabe et al. 2009*), the results of the implementation of selected mitigation measures into the calibrated model are shortly presented.

2. The earthflow phenomenon – general engineering-geological and soil-mechanical characteristics

2.1. Occurrence of earthflows with respect to geology, climate and quaternary-geological history

Geology

A large number of publications on earthflow-type landslides were analysed for descriptions of the geology assumed by the authors to be causative of these phenomena. Where these publications made no direct statement, but described the geological strata from which the affected soil covers primarily have formed, these specifications were also considered. An overview with the corresponding references is given by Table 1.

Earthflow occurrence is always bound to the outcrop of soft and deeply weathering rocks. When looking at the data from a larger number of sites, it becomes clear that nearly exclusively sedimentary or volcanosedimentary rocks are involved. In most cases, bedrock geology is made up by claystones and marls or interbedded strata containing layers of varied sedimentary rocks like claystones, shales, marls, weakly cemented sandstones or siltstones, also fine-grained volcanic tuffs or tuffites.

The relation between the dips of the bedding and the slope is postulated to play a role in earthflow development by some authors (e.g. *Bovis 1985*). However, no correlation could be found in this research. Tectonic processes clearly favour the development of such mass movements as they increase the available amount of weathering detritus, namely by the disintegration of the rocks, for example by faulting (e.g. *Corsini et al. 2001*), as well as the increase of layer thickness or outcrop area by folding or overthrusting (e.g. *Baumgartner 1981* or *Panizza et al. 1996*). The contact of the above-named rock types to dolomite or limestone formations with karst aquifers able to supply high amounts of groundwater has also been found to favour earthflow activity (e.g. *Dall'Olio et al. 1988*).

In the Lahnenwiesgraben valley, near Garmisch-Partenkirchen, Bavarian Alps, Germany, a catchment with an area of around 17 square kilometres, covering nearly the entire valley, was studied in detail with respect to the role of "gravitational mass movements in alpine sediment cascades" by *Keller 2009*. A large number of small-scale to medium-scale (max. 500 m long) shallow (< 10 m deep) earthflow phenomena can be observed in this catchment. One of these objects is the Brünstgraben earthflow, which was investigated already by *Büch 2003* (see pages B25 to B28 in the Appendix). In this work, six further earthflows in the Lahnenwiesgraben valley were mapped in detail in the field (see pages B1 to B24 in the Appendix). The above-described field work (*Keller 2009, Büch 2003,* and this work) showed that all the earthflows in that study area were related to two stratigraphic units: Kössener Schichten of the Late Triassic, consisting of interbedded strata, made up of claystones, shales, marls and limestones, and Liasfleckenmergel (Allgäu formation), Early Jurassic, made up mainly of marly claystones and weakly-bound sandstones.

Authors & year of publication	Description of location	Country	Short description of geological strata
Angeli et al. 1991	Ru di Roccia	Italy	La Valle formation
Angeli et al. 1996	Staulin, Rio Roncatto	Italy	San Cassiano formation
Angeli et al. 1998	Alverà	Italy	San Cassiano formation
Bammer 1984	Bad Goisern	Austria	Claystones and marls, Zlambachschichten, Liasfleckenmergel, Haselgebirge
Baumgartner 1981	Gschliefgraben	Austria	Claystones, clayshales, marls, sandstones, limestones; Upper Cretaceous to Early Tertiary
Berti et al. 1994	Northern Apennines	Italy	Cretaceous shales
Bertini et al. 1991	Cignale Valley	Italy	Pliocene marly clays, silty-clayey detrital cover, probably of periglacial origin
Bertolini 2001	Roncovetro	Italy	Flysch, Tertiary
Bertolini et al. 2005	Emilia Romagna	Italy	Ligurian Nappe, tectonic melanges, Late Jurassic to Early Eocene, "Argille Scagliose"
Bettelli & De Nardo 2001	Emilia Romagna	Italy	Upper Cretaceous Ligurian Flysch, Helmintoid Flysch, Ligurian Tertiary Flysch
Bockelmann 1995	Maria Eck	Germany	Flysch-Gault, Reiselsberger Sandstein varicoloured marls and sandstones
Borgatti et al. 2006	Ca'Lita	Italy	Arenitic Flysch above clay-rich rocks, Upper Cretaceous to Eocene
Bovis 1986	British Columbia	Canada	Cretaceous to early tertiary volcanics with interbedded sediments, bentonitic shales, claystones, siltstones, sandstones, conglomerates and coal
Bunza 1978	Alpbachtal	Germany	Flysch, Cretaceous marls (Cenomanmergel)
Bunza 1978	Unterammergau	Germany	Relicts of periglacial valley filling
Bunza 1978	Schwangau/Lenkerbach	Germany	Aptychenschichten
Bunza 1978	Bad Heibrunn	Germany	Upper Cretaceous Flysch, Helvetikum
Cancelli et al. 1991	Camporaghena	Italy	Scaglia rossa formation, shales, marls, limestones
Conte et al. 1991	Verbicario	Italy	Crete-Nere formation, black Flysch, Liguride unit
Conti & Tosatti 1994	Emilia Romagna	Italy	Ligurian basal complexes
Corsini et al. 1999	Corvara	Italy	La Valle formation, San Cassiano formation
Czurda & Schütz 2001	Sibratsgfäll	Austria	Clayshales, dark pelites, sandstones, conglomerates, breccias, Tertiary, Feuerstätter Decke
Dall'Olio et al. 1988	Tessina	Italy	Eocene Flysch formation, marls and claystones with intercalated calcarenites
Fischer 1967	Entlebruch	Switzerland	Flysch and subalpine Molasse, interbedded strata, mainly marls
Fleming et al. 1988 A	Manti, Utah	USA	Late Cretaceous to EarlyTertiary; Flagstaff member of Green River formation: limestones and shales deposited in freshwater environment, minor amount of sandstones and cherts; North Horn formation, fluvial and lacustrine: varicoloured clayshales and sandstones
Giusti et al. 1996	Basento Valley	Italy	Mesocenozoic structurally complex formations deriving from tectonised flysch, characterised by low sandstone/shale ratio
Herzog 1992	Neuhüttenalm	Germany	Liasfleckenmergel (Allgäu formation), Early Jurassic, marls
Herzog 1992	Erlmoosgraben	Germany	Flysch-Gault, Reiselsberger Sandstein, claystones, marls, sandstones
Keefer & Johnson 1983	San Francisco Bay Region	USA	Tertiary shales, mudstones and conglomerates; Tectonic Melange
Keller et al. 2005	Lahnenwiesgraben	Germany	Kössener Schichten, Late Triassic, marls, claystones, shales, limestones
Nozaki et al. 1994	Nagano province	Japan	Green Tuff, early Miocene; hydrothermally altered porhyrites
Panizza et al. 1996	Cortina d'Ampezzo area	Italy	San Cassiano formation
Raetzo et al. 1996	Falli Hölli	Switzerland	Gurnigelflysch, Cretaceous to Early Teriary, marly claystones, marls, sandstones
Raetzo et al. 2000	Hohberg	Switzerland	Flysch and Préalpes Médianes
Rohn 1991	Stambach (Zwerchwand)	Austria	Haselgebirge (leached salt-bearing claystones, Permotriassic), Zlambach and Allgäu formations (marls, Late Triassic to Early Jurassic)
Rohn et al. 2005	Hallstatt	Austria	Haselgebirge, Zlambach and Allgäu formations
Ronchetti 2007	Monte Modino	Italy	Flysch and Claystones
Savage et al. 1996	Slumgullion	USA	Hydrothermally altered volcanic rocks, Tertiary
v. Poschinger 1992	Inzell	Germany	Rhenodanubic Flysch, Zementmergel-Serie, interbedded strata of marls, marly and sandy limestones
Van Asch et al. 2007	La Valette	France	Flysch, Terres Noires marls, moraine deposits
Wasowski 1998	Serra dell' Acquara	Italy	Flyschoid succession of mudstones, marlstones and limestones

Table 1: Overview of rock-types involved in earthflows mentioned by different authors and for different locations.

Earthflow phenomena were also found affecting moraine deposits. But in the Lahnenwiesgraben valley, in all cases, these moraine deposits consisted predominantly of weathering material of the Kössener Schichten strata.

Climate

Earthflows are reported from all climes except for regions with year-round aridness or deserts. Wherever earthflows occur, there are at least pronounced humid seasons, i.e. in some months of the year, precipitation clearly supersedes evapotranspiration. But long-term annual averages of precipitation reported in literature for areas affected by earthflows range from very low values, e.g. 250 mm to 450 mm (earthflows of the Interior Plateau, Canada – *Bovis 1986*), or 360 mm to 1000 mm (Coast Range, California – e.g. *Keefer and Johnson 1983*), 600 mm to 1500 mm (Apennines, Italy – e.g. *Conte 1991*) up to more than 2000 mm (e.g. Vorarlberg, Austria – *Czurda and Schütz 2001*). Earthflows can also be found in tropical regions (Central Kenya – e.g. *Zoebisch and Johansson 2002*).

Quaternary-geological history

Earthflows represent an ongoing landscape- forming process. The actual earthflow activity in the alpine realm started with the retrogression of the Pleistocene glaciation (i.e. in the Lateglacial) some 15000 years ago. Radiocarbon (^{14}C) datings of wood and organic horizons (e.g. *Panizza et al. 1996, Soldati et al. 2004, Corsini et al. 2000*) inside earthflow deposits in the Cortina d'Ampezzo and Alta Badia areas (Italy) confirm that phased earthflow movements have taken place over the whole Lateglacial and Holocene period.

The data suggest, however, that earthflow activity was most extensive in two periods: At the beginning of the Holocene, mainly in the Boreal and Preboreal period, circa 11500 to 8500 years ago (cal. yr BP), when an increased amount of groundwater became available due to permafrost melting and elevated precipitation. A second cluster of earthflow-type landslides, circa 5000 to 2000 years ago (cal. yr BP), coincides approximately with the Subboreal period, which compared to the preceding Atlantic Period was characterised by increased precipitation. In the Northern Apennines, Radiocarbon data from the Emilia-Romagna region, published e.g. by *Bertolini and Pellegrini 2001*, suggest that also there, the actual earthflow activity started in the Lateglacial, showing a concentration in the Subboreal period. A study on the Holocene record of earthflow movement carried out by *Bovis 1985* in the Interior Plateau in southwest British Columbia, Canada, at ten different locations, suggested relatively modest earthflow activity during a warm and dry period covering the first 3000 years of postglacial time, a maximum of movements somewhere after 6500 years BP, correlating with the onset of cooler and moister conditions, and subsequently, a progressive decrease in the area of actively moving debris for most of the earthflows. All above-named authors mentioned that their study areas have been affected by earthflow-type slope movements also during other periods of the Holocene. After the large reactivation event at Falli Hölli, Switzerland, in 1994, *Raetzo and Lateltin 1996*, interpreting radiocarbon and dendrochronological data came to the conclusion, that in the last 5000 years, three to four large-scale reactivations had occurred, which together with the catastrophic event of 1994 indicate an average recurrence interval of around 1000 years, whereas medium-scale reactivations probably have a recurrence interval of around 400 years, and small local phases of increased activity, of less than 100 years. Based on reports in chronicles, *Baumgartner 1981*, also *Weidinger 2009* (with slightly differing interpretation of these reports), reconstructed the activity of the Gschliefgraben earthflow in Upper Austria. According to their findings, since 1600, major activity phases have occurred in 1664, 1700, 1734, 1860, 1910 and 2007. These phases were linked to larger rock fall events from adjacent mountains onto the earthflow body and/or extreme precipitation, partly also to changes in land-use.

Generally, remobilized earthflow deposits of one epoch have overridden older earthflow deposits in many places. Earthflow activity does not constitute an unexpected event (e.g. *Ratezo and Lateltin 1996*). It rather takes place in areas where in the last millennia and centuries comparable mass movements have left their deposits. The slope morphology of those areas was moulded by the earthflow movements and represents the state of a more or less unstable equilibrium.

2.2. Slope morphology and morphological features of earthflows

Earthflows occur on gently-inclined slopes and within wide depressions. Average inclinations of the earthflow bodies reported in literature range between 5° and 20°, in most cases between 10° and 15°, whereas the upper parts of the source areas are steeper. The latter areas are often characterised by relatively slow rock slide or earth slide phenomena, but also sudden rotational or translational soil-slips can locally occur in oversteepened parts of the slopes. These movements, though being of different nature, are clearly linked to the removal of material by the earthflow further downslope.

Slope inclination data of the earthflows of the Lahnenwiesgraben valley studied in this work are shown in Table 2. Morphologic features of these objects are illustrated in the maps of these slope areas in the Appendix (pages B1 to B28). The earthflows in the Lahnenwiesgraben mostly occupy depressions of bedrock topography. Where bedrock topography differs from surface topography, the extension (and also the movement) of the earthflow bodies is oblique to the slope. This could be observed more clearly in the case of the earthflow at Neuhüttenalm, near Bad Wiessee, Germany (*Herzog 1992*), which in its upper part was moving approximately diagonal to the surrounding slope.

In the Lahnenwiesgraben testing areas, it could be observed that most morphologic ridges are relatively stable and soil cover is relatively thin in these positions. While these rock ridges may become dry within a few days, water accumulates in the hollows, which due to the low permeabilities of the materials show extended wet zones over the whole year (freezing only superficially in winter), favouring deep weathering of the bedrock strata.

Earthflow area (nr. / name)	Inclination of front zone	Inclination of main earthflow area	Inclination of upper source area	Overall inclination
123	27,8°	12,0°	not clearly marked	15,6°
4	26,2°	13,8°	not clearly marked	17,2°
6	25,2°	11,6°	not clearly marked	15,8°
12 (South)	not clearly marked	11,3°	not clearly marked	11,3°
12 (North)	not clearly marked	14,4°	not clearly marked	14,4°
13	not clearly marked	10,8°	23,6°	14,3°
14	41°	14,6°	23,4°	16,3°
Brünstgraben	31°	16,9°	23,2°	18,9°

Table 2: Slope inclination of eight earthflows in the Lahnenwiesgraben valley near Garmisch-Partenkirchen, Germany.

Keefer and Johnson 1983, who carried out extensive field studies on numerous earthflows in California, developed the following scheme of an idealized single earthflow body: It starts with cracks, so-called crown cracks, followed by a main scarp and the main body of the earthflow, and ends in a bulging toe. A so-called "surface of separation" separates the earthflow toe from the overridden part of the slope. The earthflow can be divided into zones of depletion and accumulation of material. Its body is confined by lateral and basal shear surfaces and by lateral ridges in the lower part. The longitudinal profile of the earthflow surface has a sinusoidal shape. The inclination of the earthflow surface is gentler than that of the surrounding slope area.

The above-mentioned authors underline, though, that in nature, earthflows rarely appear as single earthflow bodies, but nearly always in the form of so-called "earthflow complexes" containing the deposits of several overlapping earthflow bodies. The typical morphological characteristics of such "earthflow complexes", in the following again simply referred to as "earthflows" are largely scale-independent, i.e. they are observed with very large landslides of this type, as well as with much smaller phenomena of comparable type (*Bovis 1986*).

For the alpine region, literature review and own field work showed that the morphologic appearance of slope movements of the earthflow-type can be quite different: They can occur in the form of large-scale earthflow complexes, showing deep-seated movements (10 m to 50 m depth). Average slope inclination of the main body (except for source area and front of the accumulation area) is between 5° and 15°, usually between 10° and 15°. As examples, Falli Hölli, Switzerland (*Raetzo and Lateltin 1996*), Stambach, Austria, (*Rohn 1991*), or Corvara, Italy (*Corsini et al. 2005*), may be mentioned. Furthermore, there are shallow earthflow-type slope movements (less than 10 m deep), characteristic slope inclinations are similar to those of the large-scale phenomena; lateral boundaries may be clearly distinguishable or diffuse. Several earthflows of this sort have been studied in the Lahnenwiesgraben Valley (see the maps in the Appendix, pages B1 to B24). Some shallower (< 10 m depth) earthflows are steeper (15° to 20°) and confined by bedrock topography in narrow channels, (10-50 m wide), as for example the earthflow at Neuhüttenalm, Bad Wiessee, Germany (*Herzog 1992*), but also the Brünstgraben earthflow (*Büch 2003*, see pages B25 to B28 in the Appendix). Due to the large amount of surface water converging towards the low-permeability earthflow materials in these channels, the uppermost layer of the earthflow body can sometimes pass into debris-flows.

In the following, some morphologic features of earthflow-type slope movements described by *Keefer and Johnson 1983* and *Bovis 1986*, are illustrated with examples from the earthflows studied in this work. These features are not always developed completely, but they can often be recognised in connection with such landslides.

Arcuate headscarps

Fig. 1: Headscarp area of the Brünstgraben earthflow, Garmisch-Partenkirchen, Bavarian Alps, Germany.
Tilted trees indicate rotational sliding (Photo: *Büch 2003*).

Earthflows commonly start with arcuate headscarps, which are the product of ongoing retrogression of a failure zone where rotational slides and slumps generate the earthflow material (see Figure 1). Coherent blocks get detached from the headscarp. This detachment often happens in echelon. The detached blocks may contain layers or a cover of brittle rocks that have got loosened by sliding

21

processes on the weathered soft rocks. Further downslope, the blocks break apart and get mixed up into a matrix-supported mass (e.g. *Jedlitschka 1984*).

Fig. 2: The arcuate shape of this part of the headscarp of the Brünstgraben earthflow can clearly be recognized. (Photo: *Büch 2003*).

There are different engineering-geological settings where the headscarps may be situated:

- In a thick soil cover originating from the weathering of soft rocks (e.g. Brünstgraben earthflow, see Figure 2).

- In deeply weathered and loosened soft rocks or alternating beds of soft and more brittle rocks (e.g. Ca'Lita, Northern Apennines, *Borgatti et al. 2006*).

- Where a brittle rock layer covers weathered and disjointed soft rocks (e.g. some headscarps of the Gschliefgraben earthflow, Upper Austria, *Baumgartner and Mostler 1978*)

Steep fronts of advancing lobes

Advancing earthflow lobes often show steep fronts or bulges (see Figure 3). The actively moving parts of an earthflow complex usually consist of several overlapping lobes. According to *Keefer and Johnson 1983*, the steep lobe fronts can form in three different ways:

1.) The lobe overrides the ground surface as a whole

2.) The lobe consists of one or more recumbent folds

3.) The lobe forms by imbrication of the material along thrust surfaces

The slope of the lobe front is steepened by the movements and therefore subjected to increased erosion. Eroded material spills down the lobe front and often covers its foot if the advancing of the lobe is slow enough. Vertical mixing of the material occurs as lobes thrust over those deposits. Upthrusting of the frontal part can also cause reversed slopes and closed depressions.

Fig. 3: Lobe front at the toe of the Brünstgraben earthflow. The tree trunks in the lower right part of the photograph had been driven in vertically and have been tilted (see orange dashed line) by the advancing of the lobe front.

Lateral boundary shear zones

Lateral boundary shear surfaces are observed at the boundary of the area of active movements or between areas of slower and faster movements. Where lateral boundary shear surfaces are well preserved, they are described as flat and steeply dipping (see Figure 4). Sometimes these lateral shear surfaces are slickensided. In many cases only a trace of the lateral shear surface, marked by open cracks, is visible on the ground surface.

Fig. 4: Fresh lateral boundary shear surface. Dip angle: 70°. Earthflow "Neuhüttenalm", Bavarian Alps (*Herzog 1992*).

Lateral ridges

Many earthflows show lateral ridges along the lateral boundaries of the zones affected by recent movements or zones characterised by faster movements than the surrounding areas. Well developed lateral ridges are found in areas where earthflow motion is controlled by topographic constrictions. They are absent where the movement is not convergent. There are different processes that lead to the formation of lateral ridges (*Keefer and Johnson 1983*):

1. The edge of the moving earthflow body is pushed up along a lateral shear zone, forming a so called pressure ridge (see Figure 5)
2. Earthflow material overrides the lateral boundary of the moving mass and is deposited on the adjacent unmoved ground.
3. An inactive lobe is remobilized, leaving behind its lateral parts.

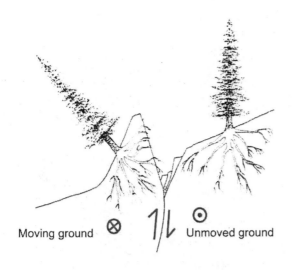

Moving ground ⊗ 7↓ ⊙ Unmoved ground

Fig. 5: Formation of a pressure ridge (modified after *Keefer and Johnson 1983)*.

Fig. 6: Remnant lateral ridges and actual lateral ridge (this ridge is a combined feature of a pressure ridge in the
upper part and a remnant lateral ridge in the lower part); track zone of the Corvara earthflow, Dolomites.

Where the actively moving surface of an earthflow complex has lowered as a result of the ongoing
downward transport of material, remnant lateral ridges (see Figure 6) can be found above the active
boundary.

Closed depressions

Closed depressions, sometimes with perennial or ephemeral ponds, form due to local reversals of the slope which are caused by the slope movements. These structures can develop between lateral ridges of different periods of movement, behind upthrusted earthflow lobes or in zones of dilatation (see Figure 7).

Fig. 7: Closed depressions with perennial ponds which have formed in a zone of dilatation; Corvara earthflow, Dolomites.

2.3. Earthflow materials

2.3.1. Soil mechanical properties

Selected soil samples and sampling areas

A literature study was performed to find out the characteristic soil mechanical properties of the materials involved in earthflow-type slope movements. In addition, 24 soil samples were taken from materials which are typically encountered in connection with earthflow activity in the alpine realm. Since most of the earthflows in the Lahnenwiesgraben testing area near Garmisch-Partenkirchen in the Bavarian Alps (Germany) are related to the Late Triasic **Kössener Schichten** strata (see Section 2.1, page 16), earthflow materials of this stratigraphic provenance were sampled. Further samples were taken of earthflow materials originating from the weathering of the Late Jurassic **Liasfleckenmergel**

(Allgäu formation) strata, as these are also involved in some of the earthflows of the Lahnenwiesgraben valley (see pages B18 and B22 in the Appendix). Materials of this stratigraphic provenance are found in connection with several earthflows in the Alps, e.g. the large Stambach earthflow near Bad Goisern, Upper Austria (*Rohn 1991*), the earthflow at Neuhüttenalm near Bad Wiessee, Bavarian Alps, and earthflow phenomena in the nearby Scheibengraben area (*Herzog 1992*). Soils that have formed from the weathering of leached (formerly salt-bearing) **Haselgebirge** claystones, Permotriassic – besides being part of the above-mentioned Stambach earthflow – were found "responsible" for numerous earthflows in the surroundings of Bad Goisern, Austria, e.g. the Sandling earthflow and two earthflows near Mount Hohe Scheibe (*Rohn et al. 2005*), as well as three earthflows in the Mount Raschberg area (*Rohn et al. 2004*). Therefore, one larger sample (10 kg) was taken from such a material. In the Dolomites, weathering material of the Triassic **La Valle and San Cassiano formations (Wengener Schichten and Kassianer Schichten)** is at the origin of several earthflows, for example the Ru di Roccia earthflow (La Valle formation, e.g. *Angeli et al. 1991*) or the Alverà, Staulin and Rio Roncatto earthflows (e.g. *Angeli et al. 1996*). These strata consist mainly of claystones, clayshales, marly limestones, sandstones and tuffites. Also the Corvara earthflow, which was chosen as a case study in this work (see Chapter 4) has formed in materials of this stratigraphic provenance.

A larger amount of samples (17) was taken at the site of Corvara, including materials from different parts of the affected slope, i.e. the Source area, Track zone and Accumulation area (see Chapter 4, pages 117 and 120 for the location of these areas). These samples were taken at different positions within the earthflow deposits, i.e. from near the ground surface and from boreholes, between shear-zones and at the depth of the shear-zones. One sample (nr. 8.331) was also taken from an older landslide deposit – most probably also of the earthflow-type – below the basal shear-zone of the actual landslide. Material originating mainly from harder volcanic components of the La Valle formation, a material which is not causative of the earthflow but can often be found embedded within the earthflow body, was also sampled (nr. 8.327). The sample was taken where this material was cut by a shear-zone and was therefore in an intensely reworked state. An overview of the samples and sampling locations is given on pages A2 to A4 in the Appendix. Unless otherwise expressly described in the text, all own classifications and tests in this Section (2.3.1.) were carried out according to the German standard DIN.

Grain-size distribution

Earthflow materials consist of a fine-grained matrix that contains a high proportion of clay and silt. This matrix encloses coarser components such as gravel, cobbles and boulders, occasionally also roots, branches and tree trunks. The earthflow material is matrix-supported, i.e. there is usually no support between the coarser components. For this reason, the soil mechanical properties of earthflow materials are mainly determined by the properties of the matrix. However, enclosed coarser components, such as cobbles or boulders, influence material behaviour; for example, they increase the overall stiffness of the materials and locally increase the shear strength.

Due to the irregular amount of the enclosed coarse components, the materials encountered in the testing areas could not be sampled representatively as a whole with respect to their grain-size distributions. At least several tons of material would have been necessary for this purpose. Consequently, only the grain-size distribution of the fraction < 2 mm was determined for each sample. The detailed results for each sample can be found in the Appendix on pages A5 to A48. The coarser components of the samples were separated and, for 16 samples, described according to their petrographic and lithological characteristics and their relative amount in this fraction of the sample (1st, 2nd...). The results of this analysis can be found in the Appendix, pages A49 to A92. The value for splitting the samples was fixed at 2 mm, because rock fragments of this size and larger were big enough so that it could easily be verified (i.e. by hand and with the naked eye) that these components were not supporting each other but were enclosed by finer material.

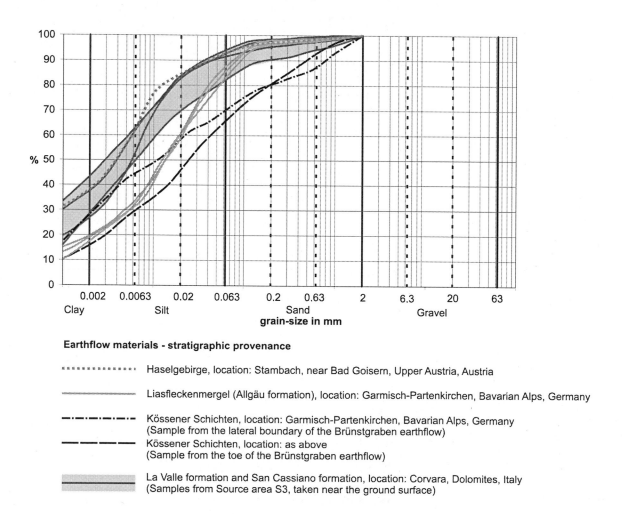

Earthflow materials - stratigraphic provenance

- ······················ Haselgebirge, location: Stambach, near Bad Goisern, Upper Austria, Austria

- ~~~~~~~~~~~~~~~ Liasfleckenmergel (Allgäu formation), location: Garmisch-Partenkirchen, Bavarian Alps, Germany

- —·—·—·—·— Kössener Schichten, location: Garmisch-Partenkirchen, Bavarian Alps, Germany
  (Sample from the lateral boundary of the Brünstgraben earthflow)
- ———— ———— Kössener Schichten, location: as above
  (Sample from the toe of the Brünstgraben earthflow)

- ▬▬▬▬▬ La Valle formation and San Cassiano formation, location: Corvara, Dolomites, Italy
  (Samples from Source area S3, taken near the ground surface)

Fig. 8: Grain-size distributions of earthflow materials (fraction < 2mm) according to their stratigraphic provenance.

The grain-size distributions of typical earthflow materials of different stratigraphic provenance (as described on pages 26 and 27) are shown in Figure 8. The grain-size distribution of the Haselgebirge soil sample from the Stambach earthflow is similar to that of the samples taken in source area S3 of the Corvara earthflow near the ground surface (see Fig. 51, page 117 for the location of this area), originating from the La Valle and San Cassiano formations. With respect to these soils, the Liasfleckenmergel samples from Garmisch-Partenkirchen are characterised by a lower proportion of clay and fine silt. As regards the Kössener Schichten samples, they both contain around 30 % of sand-size particles, i.e. considerably more than the other materials.

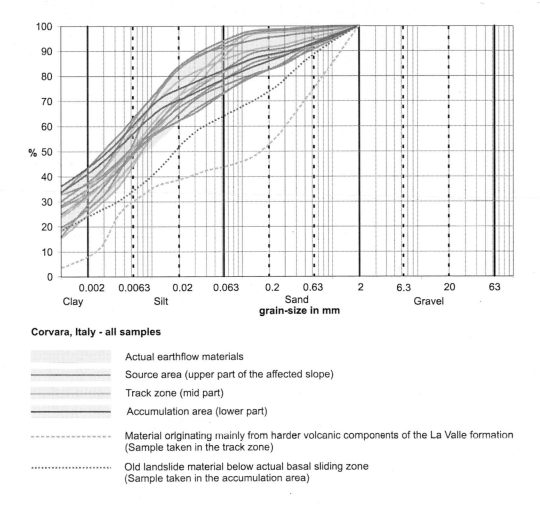

Corvara, Italy - all samples

	Actual earthflow materials
	Source area (upper part of the affected slope)
	Track zone (mid part)
	Accumulation area (lower part)
	Material originating mainly from harder volcanic components of the La Valle formation (Sample taken in the track zone)
	Old landslide material below actual basal sliding zone (Sample taken in the accumulation area)

Fig. 9: Grain-size distributions of soil samples taken in different sectors of the Corvara earthflow, Dolomites, Italy.

In Figure 9, the grain-size distributions of all samples taken at the site of Corvara are displayed. Due to the natural inhomogeneities of earthflow materials, the resulting range of grain size distributions is wider than that covered just by the smaller group of samples which were taken in source area S3 and near the ground surface. However, when considering only the actual earthflow materials (grey field in Figure 9), the shape of the envelope is similar to that of the last-mentioned group, due to the characteristic composition of the geological strata from which they originate. For example, in none of these samples, the amount of sand-size components is as high as in the Kössener Schichten samples from Garmisch-Partenkirchen, Bavarian Alps. In contrary, the grain-size distribution of the sample from the old landslide material below the actual basal sliding zone, which belongs to a landslide, presumably an earthflow, that had existed when the slope morphology and probably also the prevailing climatic conditions were different from those of today, turned out to be quite similar to that of the samples from the Brünstgraben earthflow (see Figure 8, page 28). When comparing the samples from the different sectors of the Corvara landslide, i.e. source area, track zone and accumulation area (see Chapter 4 pages 117 and 120 for the location of these areas), no significant differences in grain-size distribution could be found.

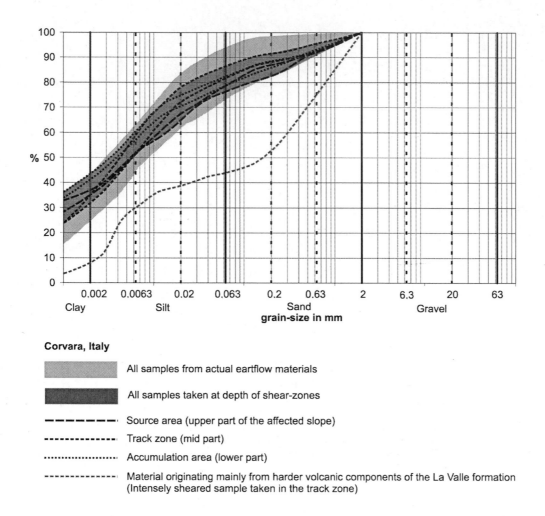

Corvara, Italy

All samples from actual eartflow materials

All samples taken at depth of shear-zones

– – – – – – Source area (upper part of the affected slope)

·············· Track zone (mid part)

················ Accumulation area (lower part)

– – – – – – Material originating mainly from harder volcanic components of the La Valle formation (Intensely sheared sample taken in the track zone)

Fig. 10: Grain-size distributions of soil samples taken at depth of shear-zones.

In Figure 10, the grain-size distributions of all samples taken in boreholes at the depth of shear-zones – indicated by inclinometer measurements or by visible signs of shearing in the borehole cores – are shown with respect to those of the whole range of grain-size distributions found at Corvara (see pages A2 to A4 in the Appendix for a detailed description of the sampling positions). Compared to the overall range of grain-size distributions of the actual earthflow materials, the samples taken at the depth of shear-zones are characterised by a relatively high proportion of clay-size particles and medium to coarse sand-size particles, while the amount of the silt and fine sand fraction is relatively low. The material originating from the harder volcanic rock components of the La Valle formation, although it was taken inside a visible shear-zone, has a completely different grain-size distribution and contains less than 10% of clay-size particles.

Description of the fraction > 2 mm

As mentioned and explained already on page 28, the grain-size distribution of the fraction >2 mm could not be determined representatively for none of the materials. In addition to the petrographic and lithological description of these materials, the rock fragments were rated according to the predominant shape of the rock fragments, their degree of rounding and degree of weathering. Results obtained by this semi-quantitative procedure were astonishingly similar for most of the samples (see pages A49 to A92 of the Appendix for the detailed results).

When comparing the samples with respect to their stratigraphic provenance, no significant differences were found, except for the observation that the samples taken from the large-scale earthflows at Corvara and Stambach contained a higher number of discernible lithologies than those from the smaller earthflows in the Lahnenwiesgraben valley. This finding is consistent with the fact that the big earthflows are able to transport the materials over larger areas. As regards the samples from the actual earthflow materials at Corvara, on average, no significant differences can be recognised between the samples taken near the ground surface, at the depth of shear-zones and between shear-zones. As only exception from this result, the average weathering degree of the rock fragments in the samples taken near the ground surface is somewhat higher than that of the rest of the samples. No significant differences were found by comparing the samples from the track zone, the source area and the accumulation area. One sample (nr. 8.330) stands out for containing 10 different rock-types. This sample was taken in the accumulation area at the depth of the actual basal shear-zone. The sampling point coincides with a former ground-surface that was exposed to fluvial and glacial deposition and later overridden by the earthflow (for a more detailed description, see pages A87 and A88 in the Appendix). The majority of the samples contained rock-types which exhibited a noticeably to extremely low slaking durability, i.e. rock fragments of these lithologies immediately started forming cracks and breaking apart – or even disintegrated to a slushy clay – when they were washed with water. Some of these rock-types could only be separated from the clay-rich matrix by crumbling before the sample was washed. These observations suggest that lithologies of low slaking durability, as they favour the fast and deep weathering processes that lead to the formation of the thick soil covers, which are required for the development of earthflows, are one of the main geological causes for these landslides.

Five of the 14 samples that were taken from actual earthflow materials at the different sites, contained organic components larger than two millimetres, such as wood or plant remains.

Plasticity and Activity

Earthflow materials from different sites and/or different regions of the world show a wide range of plasticity indexes and liquid limits, plotting along nearly the entire A-line of the plasticity chart up to liquid limits of more than 100% (e.g. *Keefer and Johnson 1983*, *Bovis 1985* or *Rohn 1991*). Figure 12 shows these values for the soil samples studied in this work. The detailed results for each sample can be found in the Appendix on pages A93 to A104.

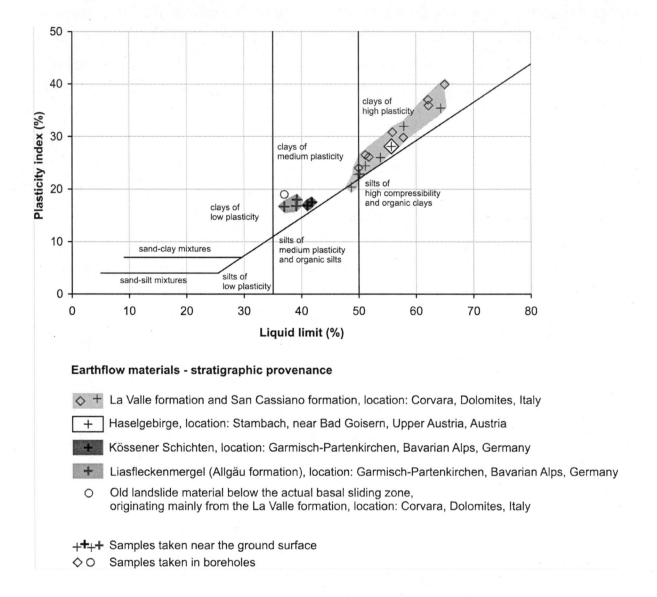

Earthflow materials - stratigraphic provenance

◇ + La Valle formation and San Cassiano formation, location: Corvara, Dolomites, Italy

+ Haselgebirge, location: Stambach, near Bad Goisern, Upper Austria, Austria

+ Kössener Schichten, location: Garmisch-Partenkirchen, Bavarian Alps, Germany

+ Liasfleckenmergel (Allgäu formation), location: Garmisch-Partenkirchen, Bavarian Alps, Germany

O Old landslide material below the actual basal sliding zone, originating mainly from the La Valle formation, location: Corvara, Dolomites, Italy

+++ Samples taken near the ground surface
◇ O Samples taken in boreholes

Fig. 11: Plasticity index and liquid limit of the studied earthflow materials according to their stratigraphic provenance.

The results presented in Figure 11 show that, when taking a larger number of samples of the same stratigraphic provenance and from the same earthflow, as in the case of the materials originating from the La Valle formation and San Cassiano formation sampled at Corvara, plasticity index and liquid limit accumulate within a wide range in the Plasticity chart, due to the natural inhomogeneities of the materials. Most of the samples of the earthflow materials from Corvara (Dolomites, Italy) are clays of high plasticity. The sample from Stambach, Upper Austria, consisting of weathering material of the Haselgebirge strata, plots in the same area, whereas the samples from the Lahnenwiesgraben Valley near Garmisch-Partenkirchen, which have formed by weathering of Kössener Schichten and Liasfleckenmergel strata, can all be classified as clays of medium plasticity. The sample taken from the old landslide material below the actual basal sliding zone of the Corvara earthflow has a lower liquid limit and plasticity index than all the samples taken from the actual earthflow materials at this site (see also the remarks on the grain-size distribution on pages 29/30).

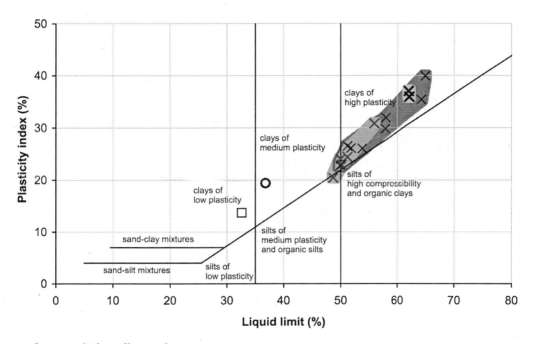

Corvara, Italy - all samples

Actual earthflow materials

⊠ Source area (upper part of the affected slope)

⊠ Track zone (mid part)

⊠ Accumulation area (lower part)

○ Old landslide material below the actual basal sliding zone
(Sample taken in the accumulation area)

□ Material originating mainly from harder volcanic components of the La Valle formation
(Sample taken in the track zone)

Fig. 12: Plasticity index and liquid limit of soil samples taken in different sectors of the Corvara earthflow.

In Figure 12 all samples from Corvara are grouped with respect to their position on the slope affected by the earthflow activity. As regards the actual earthflow materials, no significant differences can be recognised: The samples from the source area cover nearly the entire range, while the other samples, which are less in number, appear to be randomly distributed within this range. The material originating mainly from harder volcanic components of the La Valle formation, which was found embedded in the earthflow body of the track zone (but has played no active part in the development of the earthflow) shows much lower values of Plasticity than the actual earthflow materials and also lower values than the sample from the old landslide material below the actual earthflow.

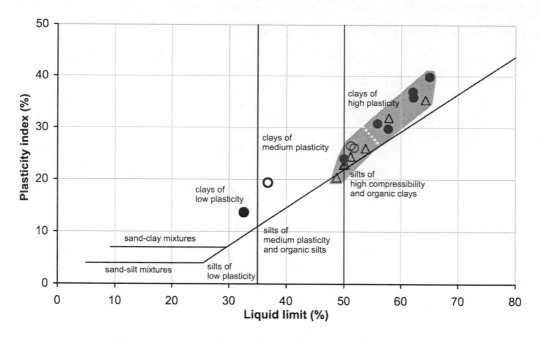

Corvara, Italy, all samples

▨   Actual earthflow materials

●   Samples taken in boreholes at depth of shear-zones

○   Samples taken in boreholes between shear-zones

△   Samples taken near the ground surface (not at shear-zones)

●   Material originating mainly from harder volcanic components of the La Valle formation (Intensely sheared sample, taken in borehole)

○   Old landslide material below actual basal sliding zone

Fig. 13: Plasticity index and liquid limit of the studied earthflow materials according to their position within the earthflow deposits.

When comparing the actual earthflow materials from Corvara with regard to their position within the earthflow deposits, it can be noticed that, on average, the samples taken at the depth of shear-zones show higher liquid limits and plasticity indexes. This finding is probably not fortuitous, because when splitting the area marked by the data from all actual earthflow materials by an arbitrary line perpendicular to the A-Line as in Figure 13, only one of the six samples taken at shear-zone depth plots within the lower part of this area, where the data of the majority of the other samples accumulate.

The activity is a measure for the influence of clay fraction on the properties of a soil and consequently the susceptibility of its properties to changes in the type of exchangeable cations and pore fluid composition (e.g. *Mitchell and Soga 2005*). As displayed in the Activity Diagram in Figure 14, most of the soil samples can be classified as clays of normal activity, some also as "inactive" (only one as "active"). The activity values of the studied soil samples show relatively large variations, between 0.6 and 1.35 when considering all samples; but also the values for samples of the same stratigraphic provenance, exhibit considerable scattering. It can be assumed that due to their parent materials, which consist of different interbedded strata, the earthflow materials contain different clay minerals in different mixing ratios.

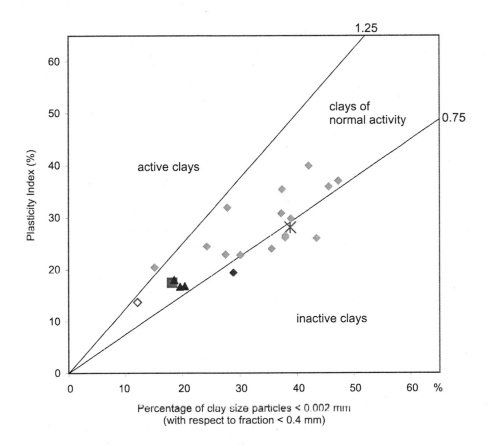

Soil samples - stratigraphic provenance

Actual earthflow materials

✳ Haselgebirge, location: Stambach, near Bad Goisern, Upper Austria, Austria

▲ Liasfleckenmergel (Allgäu formation), location: Garmisch-Partenkirchen, Bavarian Alps, Germany

■ Kössener Schichten, location: Garmisch-Partenkirchen, Bavarian Alps, Germany

◆ La Valle formation and San Cassiano formation, location: Corvara, Dolomites, Italy

Other materials from Corvara, Italy

◇ Material originating mainly from harder volcanic components of the La Valle formation

◆ Old landslide material below the actual basal sliding zone, originating from the La Valle formation

Fig. 14: Classification of the soil samples with respect to their activity.

Density of solid particles, loss on ignition and lime content

Figure 15 summarises the values obtained for density of solid particles, loss on ignition and lime content (see pages A105 to A127 in the Appendix for details on tests and test results). Density of solid particles is quite similar for the actual earthflow materials of different stratigraphic provenance. The values are normally between 2.72 and 2.76 g/cm³. A few samples (4.119, 4.122, 4.128 and 6.198) have considerably lower values, which are attributed to a small amount of organic material within these samples, since they were all taken near the ground surface (at a depth of circa 50 cm) thus next to the vegetation cover. In sample 6.198, visible wood and plant remains were found. Although loss on ignition (LOI) in this type of samples – representing mixtures of different mineralogical compositions – is the result of complex chemical processes, the high values of samples 4.119 and 4.122 (more than 6% LOI) may partly be attributed to the oxidation of organic matter.

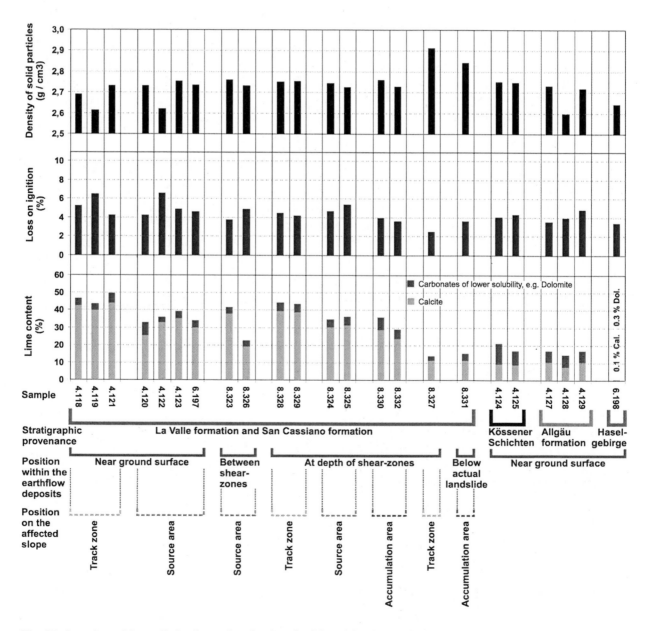

Fig. 15: Overview of the studied soil samples: density of solid particles, loss on ignition and lime content.

No significant differences can be found when comparing the density of solid particles of samples taken from the actual earthflow materials at Corvara with respect to their position within the earthflow deposits (at depth of shear-zones or not) or their position on the affected slope (Source area, Track zone, Accumulation area). Density of solid particles is highest for sample 8.327, which is no earthflow-prone material and originates mainly from harder volcanic components of the La Valle formation. The petrographic analysis of the fraction > 2 mm (see pages A49 to A92 in the Appendix) showed that the parent rock material contains mafic minerals, such as pyroxene, which are characterised by a high density. These mafic volcanic components were also found within the tuffites in the old landslide material from below the actual sliding-zone of the Corvara earthflow (sample 8.331) explaining the elevated value determined for the density of solid particles of this soil (see pages A89 and A90 in the Appendix for the petrographic description of the fraction > 2 mm).

Lime content of the actual earthflow materials from Corvara varies between 22% and 50 % and can be attributed mostly to Calcite. Sample 8.327, originating mainly from harder volcanic components of the La Valle formation, has the lowest lime content. Lime content is also reduced in the material from below the actual landslide (sample 8.331), which is characterised by a higher proportion of volcanic components than the actual earthflow materials. Lime contents of the earthflow materials from the Lahnenwiesgraben valley, which have formed from Kössener Schichten strata and from the Allgäu formation, range between 14 % and 21 %, with considerable proportions accounting to carbonates of lower solubility, such as dolomite. Lime content in the earthflow material originating from Haselgebirge strata (sampled at Stambach, Upper Austria) is very low.

Shear strength parameters

A literature review on shear strength parameters of earthflow materials reported by different authors, for example *Keefer and Johnson 1983*, *VanDine 1983*, *Bovis 1985*, *Fleming et al. 1988 B*, *Bertini et al. 1991*, *Froldi and Lunardi 1994*, *Guadagno et al. 1994*, *Nozaki et al. 1994*, *Angeli et al. 1996*, or *Bertolini, Pellegrini and Tosatti – editors – 2001* (*various authors*), gave the following results: Direct shear tests, performed under drained conditions normally give effective friction angles between 15° and 25°. Residual friction angles, obtained with different techniques, are usually between 5° and 20°. The values of the residual friction angles tend to accumulate in a margin between 10° and 16°. Values for the effective cohesion reported in literature range between 0 kPa and 100 kPa, the cohesion at residual strength is very low and may be assumed to be negligible.

The shear strength along the boundaries of the earthflow bodies is usually near the residual values as a consequence of the ongoing displacements (e.g. *Bovis 1986*). *Angeli et al. 1996*, using a ring-shear apparatus, tested two samples collected directly from the basal shear surface of the Alverà earthflow (Dolomites, Italy), which was laid open in a trial pit. Both tests gave residual friction angles of around

16°. *Rohn 1991*, who carried out drained shear tests in the triaxial apparatus on typical earthflow materials from the Stambach earthflow, Upper Austria, reported effective friction angles between 17° and 24° and effective cohesion values between 5 kPa and 20 kPa.

As regards the case studies of this work, shear-test data from the Corvara earthflow published by *Panizza et al. 2006* can be viewed in Chapter 4, Table 12 on page 151, and shear-test results from the Valoria earthflow (*Ronchetti 2009 U*) can be found in Chapter 3 Table 5 on page 62.

2.3.2. Internal structure

Only in the headscarp area, i.e. in the upper parts of the source area near the landslide crown, the sedimentary structures of the weathered source rock (parent rock) may be preserved. As a consequence of the ongoing slope movements, the main earthflow body (also earth slide body, when applying the terminology of *Cruden and Varnes 1996*) is mostly made up of reworked debris without any preferred orientation of the components (see Figure 16). Earthflow materials can therefore be regarded as inhomogeneous but more or less isotropic.

Fig. 16: Earthflow material without preferred orientation of the components. Track zone of the Corvara earthflow.

2.3.3. Discontinuities inside the earthflow materials

Instead of the primary sedimentary structures of the original rock masses (i.e. the parent rock materials), which were largely destroyed by the movements or are filled with the clay-rich material of the matrix, earthflow materials usually show a pattern of discontinuities that is mainly related to the slope movements. The following types of discontinuities are commonly observed in earthflow materials:

1.) Discontinuities caused by shear
      1a) Boundary shear surfaces (basal, lateral)
      1b) Internal shear surfaces (longitudinal)
      1c) Echelon sets of shear cracks

2.) Discontinuities caused by differential movement (tension cracks, detachment surfaces)

3) Desiccation cracks and desiccation fissures caused by shrinkage of the earthflow material.

One group of discontinuities is caused by shear. It includes lateral and basal boundary shear surfaces as well as longitudinal internal shear surfaces. The major shear surfaces are accompanied by several generations of subordinate shear cracks arranged in echelon sets. Another type of discontinuities inside the earthflow bodies is caused by divergent movements and comprises tension cracks and detachment surfaces between areas that move at different rates. Desiccation cracks and desiccation fissures caused by shrinkage of the earthflow material can pervade the uppermost part of an earthflow in times of dry weather conditions (see Figure 17). Although such desiccation cracks are not deep (usually less than one metre), they facilitate the infiltration of water from the ground surface into the earthflow body. A large number of shear-discontinuities were observed in the track zone of the Corvara earthflow (see Figure 18). Figure 19 schematically shows the pattern of discontinuities that was found by the evaluation of the average directions of the different sets of cracks that developed adjacent to the lateral boundary shear-surface. Furthermore, the qualitative horizontal velocity profile which could be derived from these observations is illustrated.

Fig. 17: Desiccation cracks, Corvara earthflow, Dolomites, Italy.

Fig. 18: Lateral boundary shear surface (left) and echelon sets of shear cracks (right), track zone of the Corvara earthflow, Dolomites, Italy.

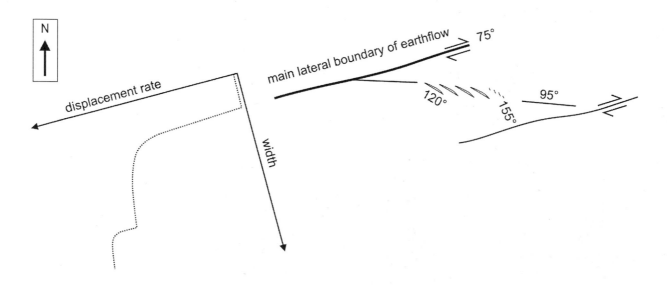

Fig. 19: Pattern of shear-discontinuities found next to the main lateral boundary shear-zone, and derived qualitative horizontal velocity profile, track zone of the Corvara earthflow, Dolomites Italy.

40

2.4. Hydrogeology of earthflows

Permeability

Laboratory tests on earthflow materials tend to give values of the coefficient of water permeability ranging between 10^{-11} m/s and 10^{-10} m/s. *Comegna 2005*, studying earthflows in the Basento valley, in the Southern Apennines of Italy, found laboratory values between 10^{-10} m/s and 10^{-12} m/s. *Ronchetti 2007*, testing soil samples from the Lezza Nuova earthflow in the Monte Modino area in the Emilia Apennines, obtained values around 10^{-11} m/s. *Angeli et al. 1991*, for samples from the Staulin earthflow near Cortina d'Ampezzo (Dolomites, Italy), report values between 10^{-10} m/s and 10^{-7} m/s.

In this work, three soil samples of earthflow materials from different sites and of different stratigraphic provenance were tested for their coefficient of water permeability: from the Corvara earthflow in the Dolomites (originating from the La Valle formation and San Cassiano formation), from the Brünstgraben earthflow in the Bavarian Alps (originating from the Kössener Schichten strata) and from earthflow number 14 in the Lahnenwiesgraben Valley, also in the Bavarian Alps (originating from the Allgäu formation). These tests gave values between $2*10^{-10}$ m/s and $3*10^{-11}$ m/s. A description of the tests and their results can be found on pages A129 to A134 in the Appendix. Maps of the Brünstgraben earthflow and earthflow number 14 are given in the Appendix on pages B21 to B28.

As the soils themselves are characterised by very low permeabilities, the in-situ permeability of earthflow materials depends on the discontinuities of the earthflow bodies and therefore is subject to great variations. **Field tests performed in boreholes on earthflow bodies** with different methods (e.g. with rising head methods) reported by different authors resulted in much higher permeability values than laboratory tests. The obtained coefficients of water permeability were usually between 10^{-6} m/s and 10^{-9} m/s. *Angeli et al. 1996*, in their field tests on earthflows of the Cortina d'Ampezzo area in the Italian Dolomites, found a typical value of 10^{-7} m/s. *Ronchetti 2007*, and *Borgatti et al. 2007 B*, summarising their results of borehole tests on the earthflows of Ca'Lita, Lezza Nuova, Tolara and Valoria (all located in the Emilia Apennines, Italy) reported values between 10^{-6} m/s and 10^{-8} m/s for the head zones, where a transition between rock slide, earth slide and earth flow phenomena was observed (i.e. the source areas). Higher values, between 10^{-5} m/s and 10^{-7} m/s, were found where very large blocks of disintegrating rock masses were intercalated within the moving materials. Values around 10^{-8} m/s were obtained for the earthflow bodies farther downslope, in the track and toe zones. *Comegna et al. 2007* report field test data from earthflows in the Basento valley (Southern Apennines, Italy), with relatively low values for the permeability of the earthflow bodies, ranging between 10^{-9} m/s and 10^{-10} m/s. Nevertheless, *Comegna 2005* and *Comegna et al. 2007*, comparing these low values with laboratory tests on materials from the same sites, found that the permeability measured in situ

41

was still one order of magnitude higher than that determined for the laboratory samples, and attributed this to a network of opened and persistent cracks and fissures present in the field.

As regards the **permeability of the rock masses in the crown zone and below the earthflow bodies**, *Ronchetti 2007* and *Borgatti et al. 2007 B* report field test values between 10^{-8} m/s and 10^{-5} m/s for the bedrock and the disintegrated rock masses in the crown zone of the Valoria, Lezza Nuova, Tolara, and Ca 'Lita earthflows (Emilia Apennines), with a frequently found value around 10^{-6} m/s. In contrast, *Comegna et al. 2007* report data from site tests on the earthflows of the Basento valley (Southern Apennines), in which values between 10^{-9} m/s and 10^{-10} m/s were obtained; and in these tests, permeabilities of the parent formation below the earthflow body were generally similar to those of the earthflow body, with the lowest values measured in the parent formation. For the Stambach earthflow near Bad Goisern (Upper Austria), *Rohn 1991* described a low-permeability earthflow body, a zone of elevated permeability at its basis, which is hydraulically connected to the ground surface via the lateral boundaries of the earthflow body, and, below this zone, a low-permeability bedrock.

Surface runoff

The hydrographic pattern on slopes affected by landslides of the earthflow type is irregular and affected by the deformation phenomena in process (e.g. *Froldi and Lunardi 1994*). In the Lahnenwiesgraben testing area, little streams often follow the course of the lateral boundaries of the earthflow bodies, the direction of lateral ridges and any type of depressions formed by the earthflow movement. At many places they spread out into wet zones, mostly marshes (see the maps in the Appendix, pages B4, B8, B12, B16, B20, B24 and B28). In a similar way, also ponds may form and disappear, but also ponds which have no surface inflow or outflow are observed on earthflows (e.g. *Williams 1988*). In the Lahnenwiesgraben valley, only very small ephemeral or perennial ponds (maximum 1-2 metres in diameter) were observed within the marshes (not displayed in the maps). The changing paths of the streams and drainage ways at (and near) the ground surface can influence the earthflow activity and vice versa (e.g. *Herzog 1992, Williams 1988*).

Subsurface hydrogeology

Investigations of different authors suggest that earthflow bodies have their own groundwater system. According to *Giusti et al. 1996,* who studied the earthflows in the Basento and Miscano valleys (Southern Apennines), a complex three-dimensional hydraulic flow arises due to the supply of water from the lateral boundaries of the earthflow body, especially if an earthflow develops in a natural drainage channel (for example in a very narrow valley). For the Alverà earthflow near Cortina d'Ampezzo (Dolomites, Italy), *Angeli et al. 1998* describe a (deeper) regional water level present throughout the earthflow body and (higher) local water levels in cracks and fissures inside the earthflow material. According to these authors, the cracks and fissures effect a rapid response of the piezometric levels measured in boreholes to precipitation and snow melt (within a few days). For the Acquara-Vadoncello earthflow, near Senerchia (Southern Apennines, Italy), *Wasowski 1998* found two different groundwater regimes, the upper related to the earthflow body, the lower related to the underlying less disturbed materials. These groundwater units were separated by a relatively impermeable horizon which was interpreted as a basal slip surface. The upper groundwater regime showed larger variations in groundwater table which *Wasowski 1998* attributed to opening and closing of cracks. Recharge times of the upper unit were estimated to be in the range of days, that of the deeper unit in the range of weeks or months.

As regards the Stambach earthflow near Bad Goisern, Upper Austria, *Rohn 1991* described the existence of a groundwater reservoir inside the earthflow body and a reservoir at the base of the earthflow mass. Pore water pressures inside the earthflow body showed only small variations throughout the year, the earthflow body was nearly completely saturated, with the unsaturated zone reaching a maximum depth of 1-2 metres. The basal reservoir, in the upper part of the slope, was characterised by pore water pressures that were significantly lower than those inside earthflow body, whereas in the lower part of the slope, pore pressures of the basal reservoir were higher than inside the earthflow body and artesian conditions were reached in the toe area of the earthflow (water outflow from boreholes was observed). The variations in pore water pressure in the basal reservoir were greater than inside the earthflow mass near the basis of the earthflow. In 5 of 6 boreholes, *Rohn 1991* found a horizon marked by small pieces of rock-fall material, wood fragments and partly also moraine gravels. The bedrock below this horizon showed an elevated porosity in the first 1-3 metres as a consequence of weathering. This region of elevated permeability corresponds to a postglacial ground surface, which was later buried by the earthflow and has a relatively fast hydraulic connection to the ground surface. A similar situation was reported by *Bovis 1985* with earthflow material acting as a confining layer to groundwater discharge in the toe area producing artesian conditions. Artesian conditions below an earthflow body were also found by other authors (e.g. *Raetzo et al. 2000*) at Hohberg, Canton Fribourg (Freiburg), Switzerland. Further aspects of the hydrogeology of earthflows, as well as additional literature references are discussed in connection with the case studies of Valoria and Corvara in Chapter 3 on pages 77 to 80 and in Chapter 4 on pages 132 to 135.

2.5. Earthflow movements

2.5.1. Velocity distribution over time – dormant stages, active stages and acceleration phases

The continuous movements of earthflows show clearly marked stages of activity (e.g. *Moser 2002*):

1.) Dormant stage

2.) Active stage (crisis event)

3) Acceleration phases

Table 3 shows some typical values of measured velocities in these different stages of activity for several earthflow-type landslides (*Moser 2002*). When regarded in the long term, earthflow movement is generally slow.

Earthflow	active stage crisis event m/d	acceleration phases without crisis m/month	dormant stage m/a	cumulative displacement crisis event (m)
Falli Hölli	6 – 7	-	0,15 – 0,5	200
Hohberg	-	0,16	0,1 – 0,6	-
Stambach	< 5	-	0,02 – 0,05	100
Gschliefgraben	-	1,8	5 – 10	-
Alvera	-	0,02	0,04 – 0,06	-
Staulin	-	-	0,02	-
Rio Roncatto	-	0,20	-	-
Apennines	-	0,02	0,002 – 0,01	-
Pavilion	-	0,07	0,07 – 0,3	-
La Lécherette	-	-	0,1 – 0,4	-
La Frasse	-	lower zone: 1 - 2	upper zone: 0 – 0,1 medium zone: 0,1 – 0,15 lower zone: 0,15 – 0,6	-
Handlova	6	-	-	240
Lubietova	6,5	-	-	50
Slumgullion	-	central zone: 1,0	toe zone: 0,75 central zone : 6,0	-
Manti	3,0	-	0,04	90

Table 3: Velocities at different stages of earthflow activity for several earthflows (*Moser 2002* and references cited therein).

Dormant stage

Movement rates of the dormant stage reported from different sites are quite different and range from centimetres per year up to metres per year (see Table 3). For the earthflows in the Lahnenwiesgraben

44

valley near Garmisch-Partenkirchen, Bavarian Alps, which between 2000 and 2006 were all in a dormant stage, precision tape measurements (*Keller 2009* and further measurements carried out in the context of this thesis) gave displacement rates between some millimetres per year and a few decimetres per year (see Table 4). Interpretation of the one-dimensional data obtained from the tape measurements in order to estimate the displacement rates of the earthflow bodies, was based on the 1:1000 field maps shown on pages B1 to B28 in the Appendix.

Earthflow (nr. / name)	Range of displacements per year
123	no data
4	mm
6	mm
12 (South)	cm - dm
12 (North)	cm
13	cm - dm
14	cm
Brünstgraben	dm

Table 4: Range of displacements per year for eight dormant earthflows in the Lahnenwiesgraben Valley near Garmisch-Partenkirchen, Bavarian Alps, Germany (*Keller 2009* and this study).

Regarding the case studies investigated in Chapters 3 and 4, the Corvara earthflow is at present in a dormant stage and movement rates are typically in the range of decimetres per year, in some parts of the narrowed track zone also some metres per year were measured (e.g. *Corsini et al. 2005, Panizza et al. 2006*). The behaviour of the Valoria earthflow was characterised by several active stages since 2001, for earlier times no measurement data are available. For this reason, it depends on the interpretation of the data, which movement rates can typically be attributed to the dormant stage and do not represent elevated rates that are preceding or following reactivation events. Characteristic rates of the dormant stage are most probably between mm per year and cm per year (*Ronchetti, F., University of Modena and Reggio Emilia, personal communication*).

Active stage

Movements of the active stage of earthflows are in the order of metres per day, with total displacements reaching up to more than 200 metres (e.g. *Bonnard et al. 1995, Raetzo and Lateltin 1996*). In some cases, also displacements of tens of meters per day have been observed during the active stage (e.g. *Poschinger, v., 1992*). The high activity can last for some days or weeks and then suddenly stops, i.e. movements quite abruptly start decelerating until reaching the low rates of the dormant stage.

Data recorded continuously and in short measurement intervals during such active stages is quite rare in literature, because often measurement devices or benchmarks get destroyed by the large deformations. The following movement pattern is reported in *Keefer and Johnson 1983,* for an earthflow near Minnis North, Northern Ireland. This movement pattern consists of a succession of major surges, i.e. during a few minutes, the earthflow surged forward several metres at a velocity in the order of metres per minute. Most Surges began suddenly and were followed by periods of smooth gradual deceleration. Altogether, the resulting displacements over several days were in the range of metres or tens of metres. However, due to the above-mentioned scarceness of continuous data and high-frequency data, it is not clear whether all earthflows are able to move in such surges.

Accelerations without crisis events, i.e. without reaching the active stage

The short-term behaviour of earthflows is sometimes characterised by sudden accelerations without crisis events, i.e. without reaching the high movement rates of the active stage. Displacement rates in these acceleration phases are between two and 60 times higher than those of the dormant stage (*Moser 2002*). Velocities reported in literature for such phases range from cm per month to m per month

Keefer and Johnson 1983, reviewing measurement data of other authors which were using continuously recording devices for investigating earthflows that moved at relatively persistent rates in the range of metres per month, reported that there were two movement patterns resulting in this type of displacements:

a) Movement at relatively constant velocities

b) Slip-stick movement

In the first case, the earthflow is moving at a constant velocity showing periods of moderate acceleration and deceleration. In case b) the earthflow advances at low velocities for a certain time (a few hours) and then abruptly surges forward a few millimetres or centimetres. The displacement rate is determined by the number of these small surges within a given period.

2.5.2. Factors controlling and triggering earthflow movement

Factors controlling earthflow movement in the dormant stage

A correlation between meteorological conditions, water levels and displacement rates was found by many authors. As a matter of fact, earthflow movement in the dormant stage appears to be controlled by climatic conditions, that is, especially by rainfall and snowmelt recharge of the groundwater reservoir. For example, *Giusti et al. 1996* stressed that for one of their case studies in the Southern Apennines of Italy, it was possible to evaluate a threshold value for the pore water pressure below which displacement rates should be nearly zero. *VanDine 1983,* for the Drynoch earthflow in British Columbia, Canada, reports a correlation between spring runoff, high groundwater levels and maximum movement rates and further relates: During the construction works for the Trans-Canada Highway in the toe zone of this large earthflow, drilling engineers observed movements increasing locally within one day following commencement of drilling with water on Monday and decreasing within two days after cessation of drilling on Friday.

Measurements of displacements of the shallow earthflow phenomena in the Lahnenwiesgraben testing area have shown that these dormant earthflows move at relatively constant and persistent rates, showing seasonal acceleration phases during summerly wet periods, which are characterised both by a high frequency and a high intensity of rainfalls. Figure 20 displays the typical course of displacements over time measured for one of these earthflows.

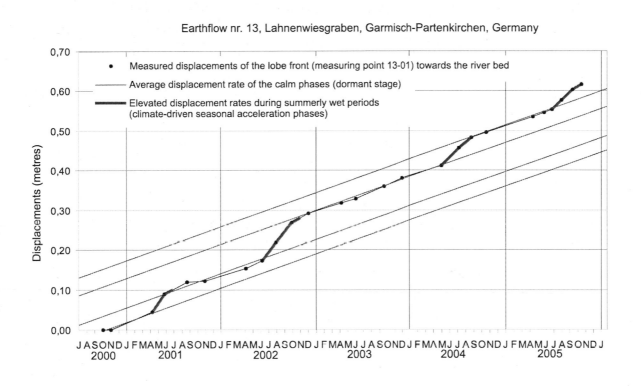

Fig. 20: Course of displacements over time determined for the toe of earthflow nr. 13, Lahnenwiesgraben valley, Bavarian Alps (*Keller 2009* and this study).

Between 2000 and 2005 the toe of earthflow nr. 13 has advanced around 60 cm. (Maps of this earthflow can be viewed in the Appendix on pages B17 to B20). A displacement rate of around 8 cm per year has been the more or less constant "background" velocity of this movement, whereas significantly higher movement rates were observed during summerly wet periods (compare Figure 21), which are typical for the climate in the Lahnenwiesgraben valley. In the year 2003, which was characterised by a dry summer all over Germany (and also beyond), there was no summerly wet period (*Keller 2009*) and consequently, no seasonal acceleration of the movements could be measured. For earthflow number 13, *Keller 2009*, based on the interpretation of inclinometer, borehole extensometer and surface displacement measurements, found that deformations concentrated along a shear-zone of less than one metre thickness, located at a depth of around four metres.

Figure 21 shows measurements of pore pressure inside the earthflow body carried out by *Meier, H., 2005 U* using an open pipe piezometer with a fissured tract covering the first four meters below the ground surface, i.e. until the detected shear-zone. In addition to the measured course of pore pressure, precipitation data from a meteorological station at circa 500 metres distance (*Haas 2005 U*), and the displacement data from Figure 20 are illustrated.

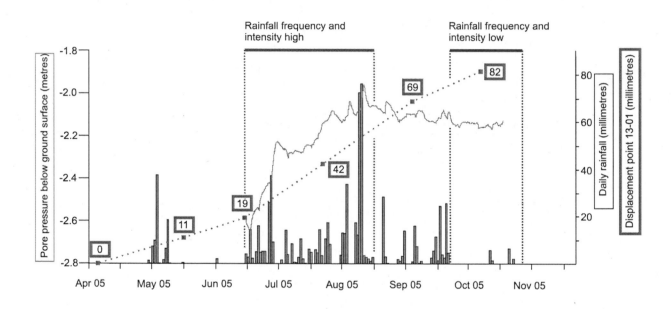

Fig. 21: Daily rainfall, pore pressure and displacements for earthflow nr. 13, Lahnenwiesgraben valley, Bavarian Alps, between April and November 2005 (data from *Haas 2005 U, Meier, H., 2005 U*, and this study).

The data show that in the middle of June 2005, a period of high rainfall frequency and intensity had started. Three days after this change of the meteorological conditions, pore pressure in the borehole started rising. By the middle of July, intensity of the rainfalls decreased, accompanied – with some days delay – by a slight decrease of measured pore pressure. The amount of rainfall re-increased in August and decreased a bit in September, before dry weather conditions were again observed after the 6th of October. The trend of measured pore pressures followed this distribution of precipitations, within a few days. During October and the first days of November, measured pore pressures remained still quite high. This is because in addition to the relatively fast infiltration of precipitation and near-surface runoff waters through high-permeability discontinuities, there is also a continuous and slow infiltration of water (and pore pressure) from the hydrographic pattern into the low-permeability earthflow body – especially from the extended wet zones, marshes and little ponds. (Please, see the maps of earthflow nr. 13 on pages B17 to B20 in the Appendix). These reservoirs are present during the whole year, but they get replenished gradually by snowmelt and by the following summerly wet periods, and then diminish very slowly during autumn, when usually precipitation decreases, but evapotranspiration is low, and further diminish during the winter, when infiltration is restricted by frost and snow cover. As it can be observed in Figure 21, the trend of the displacement rate clearly follows the seasonal course of pore pressures, i.e. displacement rate starts to increase somewhere between the beginning of July and the beginning of August, and it starts decreasing between the middle of September and the beginning of October. Unfortunately, no further displacement measurements could be performed in November, as a relatively early and thick snow cover had made the measurement points inaccessible.

Based on the data gathered on the earthflows of the Lahnenwiesgraben valley, the following conclusions were drawn:

　　1.) All the earthflows in this testing area are at present in the dormant stage.

　　2.) In this stage, their movements are caused by slow but ongoing creep deformations
　　　　which occur mainly along shear-zones and take place under drained conditions.

　　3.) The speed of these slope movements, i.e. the rate of the creep deformations,
　　　　is controlled by the level of pore water pressures in the slope.

　　4.) Variations of these pore water pressures – and consequently also of the displacement rates–
　　　　are moderate and associated with seasonal changes in the amount of infiltration.

Factors causing the transition to the active stage

The results of several authors showed that for the transition of earthflow movement from the very low displacement rates of the dormant stage to the much higher rates of the active stage excess pore pressures play a very important role. Excess pore pressures caused by undrained loading were proved to be a fundamental mechanism for earthflows [mudslides] already by *Hutchinson and Bhandari 1971*. *Comegna et al. 2007,* studied earthflows [mudslides] in the Basento valley in the Southern Apennines (Italy), particularly the Masseria Marino earthflow, which was characterised by periods of high activity (movement rates of metres per day to tens of metres per day) and periods of very low activity (cm per year). *Comegna et al. 2007*, analysing the data of a monitoring period that lasted ten years and comprehensive field data that had been gathered on these earthflows since the year 1979 (e.g. *Giusti et al. 1996*) came to the conclusion that movements were governed by alternating phases of undrained and drained deformation processes. By means of numerical analyses, *Comegna et al. 2007* were able to show that "any redistribution of the internal state of stress associated with local mobilisation of the mudslide [earthflow] body" can induce such excess pore pressures. The following scenarios are most often found in literature in connection with drastic accelerations of earthflow movements:

a) A gradual rise of the pore water pressure inside or below the earthflow body up to a critical value.

b) Loading processes, e.g. head loading, producing both an additional driving force and (rapidly) causing excess pore water pressures (undrained effect).

c) A rapid change in stress situation induced by earthquakes.

In **scenario a)**, acceleration most probably starts under drained conditions, i.e. without creation of significant excess pore pressures. A critical pore pressure situation is reached gradually, e.g. by adverse climatic conditions lasting over extended periods. This rise of the pore water pressure can cause a drastic acceleration of the movements along shear-zones (see Figure 22), or the remobilisation of inactive shear-zones. If these movements become fast enough so that pore pressures generated during the deformation of the soil do not have enough time to dissipate, then excess pore pressures build up, which cause a further acceleration, finally resulting in the high displacement rates of the active stage (see page 45). There is quite clear evidence that the above-described scenario was responsible for the reactivation of the Falli Hölli earthflow, Canton Fribourg, Switzerland, in 1994: *Bonnard et al. 1995* found that precipitation recorded by two meteorological stations in the area of the Falli Hölli earthflow in the 17 years before its active stage in summer 1994 was 14 % and 34 % above the long-term average. Furthermore, 2750 mm of rainfall were measured between June 1993 and May 1994. In addition, *Bonnard et al. 1995* stressed that the winter 1993/1994 was characterised by the exceptional occurrence of three phases of snow melt. A photogrammetric study carried out after the event of 1994 revealed that between 1981 and 1987, as well as between 1987 and 1993, considerable slope movements had already taken place in an area located circa 500 metres upslope of the village of Falli Hölli. The maximum total displacements that were detected in this way were around four metres with displacements in the first interval locally exceeding one metre (*VERSINCLIM*).

In the Lahnenwiesgraben testing area, between 2000 and 2006, the Brünstgraben earthflow moved at an average velocity of some decimetres per year (*Keller 2009* and this study). In summer 2002, extreme precipitation and extreme surface runoff, especially in the lower part of the channel occupied by the earthflow (see the maps on pages B25 to B28 in the Appendix) caused a drastic acceleration of the movements (see Figure 22). Although displacements were not recorded continuously, it can be inferred that the rates reached at least several decimetres per month, in the region downslope from measuring point 08-00 (see the map on page B28 in the Appendix, for the locations of the measuring points). However, the movements slowed down in autumn, without having reached a rate typical of an active stage. The acceleration phase of 2002 can therefore be classified as described on page 46. Either the deformation rates reached in connection with this acceleration phase were still too low to cause significant excess pore pressures, or the excess pore pressures were too small for bringing about a further acceleration of the movements.

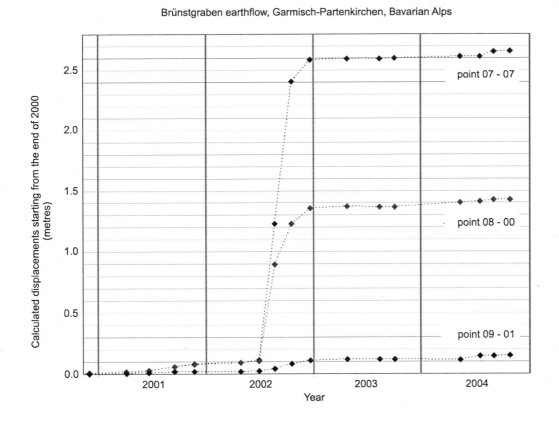

Fig. 22: Course of displacements with time for three measuring points on the Brünstgraben earthflow, Garmisch-Partenkirchen, Bavarian Alps.

The loading processes of **scenario b)** may be consequences of an acceleration of earthflow movements as described by scenario a) – for example if an advancing earthflow lobe deposits material or exerts a thrust onto the earthflow body located downslope – or loading is caused by other types of slope movements. These may be faster phenomena inside the earthflow area, such as a local collapse or failure of oversteepened zones in the headscarp region. But also slope movements outside the earthflow can deposit landslide material next to or onto the earthflow body and cause loading. For example, *Fleming et al. 1988 A and B* stated that the large Manti earthflow in Utah (USA) was reactivated in the spring of 1974 by the loading of a debris flow from the rim of Manti Canyon. Debris flow deposits overlapped with the source area of the earthflow. Movement propagated downslope during the next 12 months such that in the summer and fall of 1975, an earthflow mass about 3 km long was moving at a rate of as much as 3 m per day. Another example for scenario b) was described by *Rohn 1991* for the Stambach earthflow near Bad Goisern in Upper Austria. The active stage of this earthflow in 1982 was preceded by a series of three major rock-fall events, each of them had deposited several ten thousands of cubic meters onto the source area of the dormant earthflow. According to

Rohn et al. 2004, who studied large-scale lateral spreading of rock masses in the Hallstatt zones of the Northern Calcareous Alps, the geological situation of a rigid rock slab (e.g. limestones or dolomites) overlying a ductile substratum (e.g. claystones) favours the reactivations of earthflows in the context of the above-mentioned scenario b). The main processes were described by these authors in the following sequence: lateral spreading, toppling, rock fall, undrained loading, earthflow activation.

As regards **scenario c)**, *Panizza et al. 1996,* reported that the earthquake on 15th of September 1976 in the Friuli / Venezia Giulia Region of Northern Italy triggered an earthflow near Cinque Torri. After the Irpinia earthquake with a magnitude of 6.9, affecting the Southern Apennines in 1980, several earthflows were remobilised, e.g. the large Serra dell'Acquara earthflow near the town of Senerchia (e.g. *Wasowski 1998*), and some of these earthflows moved many tens of meters during a few days (*Giusti et al. 1996*). *Comegna et al. 2007* stated that the large earthflows triggered by this earthquake were caused by excess pore pressures that had been induced due to the "effects of seismic (cyclic) loading".

2.5.3. Spatial velocity distribution

Longitudinal velocity distribution at the ground surface

Several authors observed displacements of different magnitude along earthflow bodies and recognised zones of extension and compression, showing that the earthflows as a whole did not behave as a rigid bodies (e. g. *Giusti et al. 1996* or *Pellegrino et al. 2000* for the earthflows in the Basento valley, Southern Apennines, Italy). Also for the huge Slumgullion earthflow in South-western Colorado (USA), *Savage et al. 1996* described a zone of compression in the lower part and a zone of extension in the upper part of the affected slope.

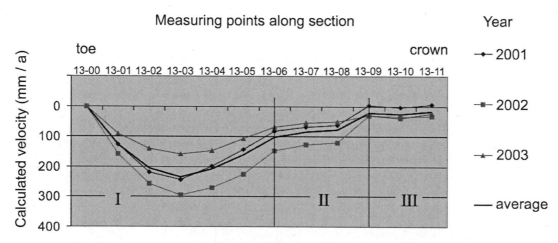

Fig. 23: Results of displacement measurements on earthflow nr. 13, Garmisch-Partenkirchen, Bavarian Alps (modified after *Keller 2009*).

Similar observations could be made also in the case of the much smaller earthflows in the Lahnenwiesgraben valley, e.g. earthflow nr. 13 (see pages B17 to B20 in the Appendix). Figure 23 shows surface displacement rates of measuring points along a longitudinal section of earthflow nr. 13 calculated in relation to the immobile checkpoint 13-00 (see page B20 in the Appendix for the locations of these measuring points). As can be seen in Figure 23, there are several zones characterised by different movement patterns. The points located in zone III in the uppermost part of the affected slope are moving at the lowest rates, whereas zone II moves at a higher rate and seems to separate from zone III along a zone of extension (not denominated in Figure 23). Within zones II and III, displacement rates of the neighbouring measuring points are more or less equal, this means that no significant extension or compression occurs. On the contrary, zone I is characterised by a gradual transition from an upslope zone of extension to a downslope zone of compression.

Horizontal velocity distribution at the ground surface

Keefer and Johnson 1983, also *Bovis 1986*, based on measurements on lines of survey stakes, pegs or fence posts and a review of similar data published by other authors, found that the bulk of surface displacements occurs at lateral shear-zones, rather than within the moving mass. However, according to *Keefer and Johnson 1983,* some internal deformation also occurred due to shear on discrete internal (longitudinal) shear-zones, and some earthflows also revealed a significant component of distributed shear or flow. For the Stambach earthflow near Bad Goisern, Upper Austria, *Rohn 1991* described that in addition to the movements along the lateral boundary shear-zones, also internal deformations were observed, as forest roads were deflected to curved shapes or sheared along internal shear-zones.

Velocity distribution over depth

Figure 24 presents schematic pictures of typical inclinometer profiles reported for earthflows in literature. Profile types A, B and C were abstracted by *Froldi and Lunardi 1994* from a large measurement campaign with 38 inclinometers on nine different sites in the North-eastern Apennines of Italy. **Type C**, was the most frequent deformation pattern found during literature review. In this case deformations concentrate in a basal shear-zone and deformations of the rest of the earthflow body are very small. Such inclinometer profiles were found for example at the Staulin earthflow and the Rio Roncatto earthflow in the Cortina d'Ampezzo area in the Italian Dolomites (e.g. *Angeli et al. 1996 A*), at the Acquara-Vadoncello earthflow near Senerchia in the Southern Apennines (*Wasowski 1998*), or at the Pavilion earthflow in Southwest Britisch Columbia, Canada (*Bovis 1986*).

In the Lahnenwiesgraben testing area, where only one inclinometer profile, in earthflow number 13, was measured, the obtained deformation profile was also of type C, i.e. characterised by one basal shear-zone (*Keller 2009*).

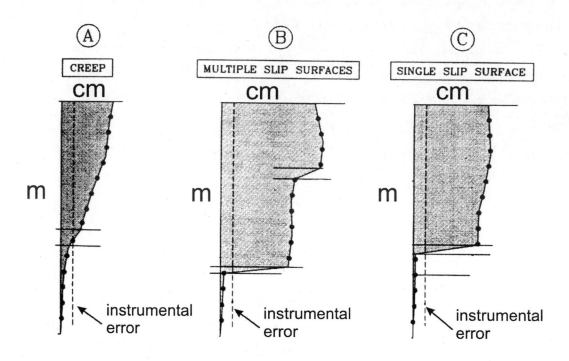

Fig. 24: Typical inclinometer profiles of earthflow-type slope movements (slightly modified after *Froldi and Lunardi 1994*).

Type B was for example reported by *D'Elia et al. 2000*, who investigated five earthflows along the Inonic Coast near Capo Spulico in South-eastern Italy. Also in the case of the Corvara earthflow the inclinometer measurements indicated secondary shear-zones in boreholes C4 and C6 (*Panizza et al. 2006*; see pages A3 to A4 in the Appendix for illustration). However, these were characterised by very small amounts of deformation compared to the basal shear-zone, so that these deformation profiles may be regarded as the transition between the cases B and C.

Profile **type A** is characterised by a continuously increasing amount of displacements towards the ground surface, which starts at a certain depth, but no shear-zone can be localised. This type is commonly associated with the slowest movements (mm per year, maximum very few cm per year). *Haefeli 1967* termed this type of deformations as "continuous creep / Kontinuierliches Kriechen". Typical examples for type A were found by three inclinometers installed at the dormant Stambach earthflow near Bad Goisern, Upper Austria, six years after its last active stage in 1982. Rates of displacements measured by *Rohn 1991* were around 5 mm/year in two of these boreholes and 3 mm/year in the third borehole. Displacements started at depths between 19 m and 26 m. According to *Rohn 1991*, in this situation, stresses were below the residual strength of the earthflow material and the basal shear-zones had become inactive.

Virtually all the data available on the distribution of the movements over depth was obtained with inclinometers. For this reason, data are only obtained as long as movement rates are small. As soon as earthflow movements accelerate, the inclinometer tubes become unusable as a consequence of the deformations. This means that practically no measured data is available on the subsurface deformation pattern during acceleration phases and active stages of earthflows.

Some inclinometer profiles that were obtained shortly before the destruction of the tubes showed rather strong internal deformations over the entire depth of the moving mass, for example profile S1' in Figure 25, measured in the Masseria Marino earthflow in the Basento valley (Southern Apennines). According to *Comegna et al. 2007,* it seems that in this case, the most active parts of the earthflow body experienced large strains over its entire thickness, whereas in the less active parts, shear strains concentrated in the shear-zone. *Comegna et al. 2007* further stated that probably only in very active stages of movements characterised by the spreading of high pore pressures (excess pore pressures) within the entire thickness of the earthflow body, shear strains become significant as shown by Inclinometer profile S1' in Figure 25.

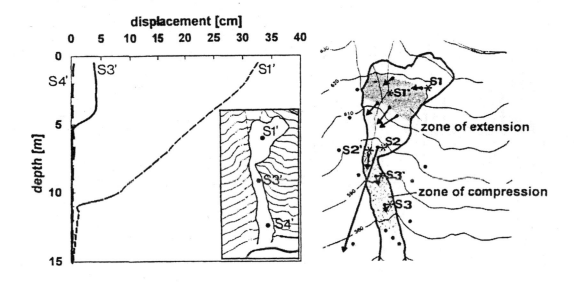

Fig. 25: Inclinometer profiles measured at the Masseria Marino earthflow, Basento Valley, Southern Apennines, Italy (*Giusti et al. 1996*).

At the large Falli Hölli earthflow in the Canton Fribourg (Switzerland), one inclinometer (Chlöw 3, *VERSINCLIM 1998*), exhibited a similar deformation pattern: At first, the measurements showed a basal shear-zone at a depth of 23 metres. The region below this depth was not affected by the movements and the region above was moving more or less en bloc. When movements accelerated, displacements were distributed approximately uniformly over the whole range between zero and 23 metres depth. Also in this case, this type of inclinometer profile was most probably associated with high pore water pressures (excess pore pressures), as the authors of the above-cited report describe a rise of the water level in all fissures of the earthflow mass until the ground surface, visible in the field.

As in such cases, the deformations are near the limit of the measuring method, the inclinometer data usually do not allow for determining, whether these profiles result from deformations that affect the entire mass rather continuously (as in type A) or from a larger number of shear-zones (as in type B). As the earthflow materials studied in this work are quite inhomogeneous and have numerous discontinuities, it can be presumed that the reduced shear strength would at first be exceeded along existing discontinuities or weak zones within the mass, this means that deformations are more likely to concentrate along multiple internal shear-zones.

In the active stage of many earthflows, the material near the ground surface becomes so soft that it does not support a person's weight (e.g. *Keefer and Johnson 1983*). In such cases it seems more likely that this material deforms as a whole, but it is not clear if the soil reaches such a soft consistency also at greater depths.

3. Reconstruction and numerical modelling of past evolution phases in the source area of the Valoria earthflow

3.1. Short Introduction to the Valoria case study

The Valoria earthflow is located in the Emilia Apennines, in the upper Secchia River basin. It affects an area of 1.6 km² over a length of about 3.5 km (see Fig. 26). At present, this landslide can be classified as a complex rock slide earth slide earth flow (after *Cruden and Varnes 1996*). In the last sixty years, until January 2008, the Valoria earthflow has been reactivated eight times (1950, 1951, 1956, 1972, 1984, 2001, 2005 and 2007). New reactivations followed and the related processes are going on at present. During these critical events, the morphology of the landslide head area has undergone major changes (*Ronchetti 2007, Ronchetti 2008 U*).

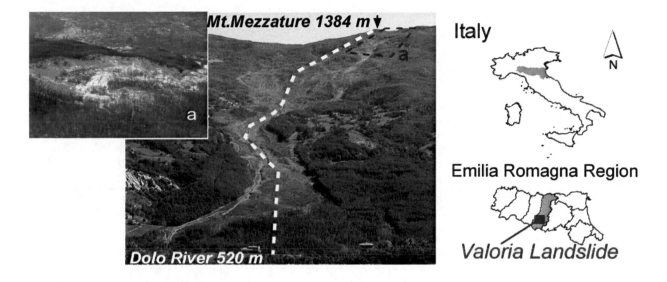

Fig. 26: Geographic overview of the Valoria earthflow, displaying the location of the studied slope area and the longitudinal section of Fig. 27 (from *Ronchetti 2008 U*).

The Valoria landslide can (at present) be subdivided into the following different zones (see Fig. 27). In the head and crown zone, the **upper source area** is characterised mainly by **rock slide and earth slide** phenomena and extends between 1375 m and 1200 m above sea level. This area is marked by the letter "a" in Fig. 26 and represents the object of investigation in this research. There, rotational and translational movements involve clayey and flysch-type rock masses outcropping downwards from the crown and the deposits resulting from the weathering and deformation of these. In the **lower source area**, which is located between 1200 and 925 m, where the displaced rock masses and earth-slide deposits are completely dismembered and then incorporated into several **earth flows**, converging into a single earth-flow track. This **earth-flow track** extends from 925 to 650 m. The material is then accumulated along an **earth-flow toe** which is located between 650 and 520 m and finally cut by the Dolo River (*Ronchetti 2007, Ronchetti 2008 U*).

57

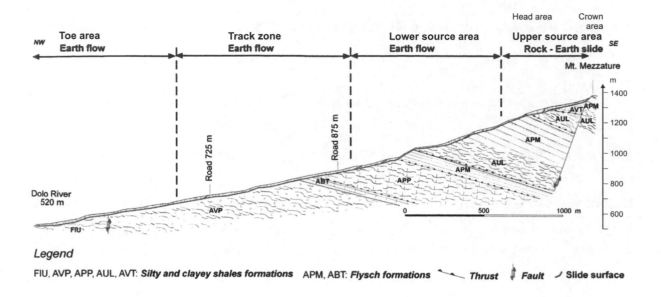

Fig. 27: Longitudinal section of the landslide showing the different bedrock formations and landslide-zones mentioned in this work (slightly modified after *Ronchetti 2007*). See Fig. 26 for the location of the section.

Drill holes and refraction seismics (*Baldi et al. 2009*) have shown that the thickness of the sliding rock and soil masses in the area investigated in this work is between five metres above the steeper parts built up by clayey shales and sandstones and up to about 35 metres in the flat zone covering the most clay-rich bedrocks. At present, the latter area is undergoing major erosion. Farther downslope, the thickness of the earthflow deposits along the slope varies from a few metres in high slope-gradient regions to more than 30 metres in the low slope-gradient regions of the track and toe areas, where deposits of different ages have accumulated (*Ronchetti 2007*).

3.2. Geological setting

The Northern Apennines are a fold-and-thrust belt that has formed during the Tertiary by the convergence of the European and Adriatic plates *(Bettelli and De Nardo 2001)*. The Valoria landslide affects Cretaceous to Miocene rock masses such as sandstone-dominated flysch and silty to clayey shales (see Fig. 28). The slope part investigated in this work is built up mainly by the silty clayshales formation "Argilliti dell' Uccelliera" and by the claystones formation "Argille variegate di Grizzana Morandi". The steeper slope area in the lower part consists mainly of flysch-type weakly bound sandstones of the "Arenarie del Poggio di Mezzature" formation. Their higher erosional resistance has led to a step-shaped morphology, which is also observed in the other zones of the Valoria landslide (*Ronchetti 2007*).

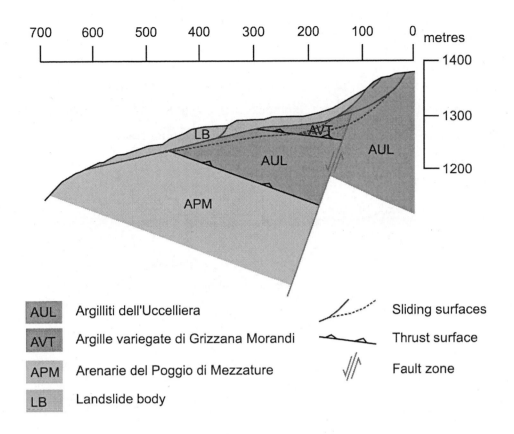

AUL	Argilliti dell'Uccelliera	
AVT	Argille variegate di Grizzana Morandi	
APM	Arenarie del Poggio di Mezzature	
LB	Landslide body	

Sliding surfaces

Thrust surface

Fault zone

Fig. 28: Geological profile through the studied slope along the modelled section (drawn after *Ronchetti 2007* and *Ronchetti 2008 U*).

All these formations are characterised by a pronounced interbedding of quite different strata. The "Arenarie del Poggio di Mezzature" formation comprises thick-bedded coarse-grained turbiditic sequences as well as thin-bedded and fine-grained ones, and different types of pelitic sediments lie between the turbidites (as reported e.g. by *Ronchetti 2007*). The "Argilliti dell' Uccelliera" and "Argille variegate di Grizzana Morandi" formations are made up mainly of different types of silty clayshales and claystones but contain also layers of siltstones, marls and limestones, and intercalations of flysch-type sandstones. In the mostly red-coloured clayshales of the "Argilliti dell' Uccelliera" formation, also embedded nodules of pyrite and plant rests substituted by native copper have been found (as reported e.g. by *Ronchetti 2007*).

Inside the slope, the rocks are deformed by overthrusts and faults. These tectonic processes formed a high number of pre-marked discontinuities and weak zones favouring the weathering of the rock masses, which anyway are susceptible to fast weathering because of their geological composition. In this way, the tectonic processes have favoured also the development of the landslide. As a consequence, the first detachment zone of the landslide that affected the studied slope area has formed near an important normal fault where due to an elevated density of the discontinuities the rock mass is weaker than in the surroundings. However, as it can be seen in Fig. 28, the landslide activity is not controlled by the directions of the main structural features. The overthrust surfaces which are more or less parallel to the bedding planes, as well as the steep normal fault, are dipping into the slope. As a result, in the crown and head zone, the landslide has developed along listric shear-surfaces formed by failure through the highly fractured rock masses. Therefore, at the scale of interest in this study, approximately isotropic behaviour involving failure on closely spaced discontinuities is taken into consideration for the geotechnical classification of the rock masses in Section 3.3., instead of considering the behaviour of blocks and wedges defined by intersecting structural features.

The landslide body is made up of earth-slide material that can be described as rock blocks in a silty-clayey matrix. Deeper inside the slope, the landslide affects also, to a minor degree, the weathered rocks, disintegrated into blocks in the scale of metres, which – same as their parent bedrock – consist of different types of clayshales and weak sandstones (*Ronchetti 2007*). Probably a large part of the fine-grained fractions of the earth-slide material have evolved from a thick residual soil and from colluvium accumulated on top of it in a slightly concave slope morphology (regarding the transversal slope profile, perpendicular to the studied section). Because of the large deformations that have occurred during the critical landslide-events, these layers cannot be distinguished any more.

3.3. Geotechnical properties of the materials

As previously described, the bedrock in the head and crown area of the Valoria landslide is characterised by clayey and silty as well as flysch-type sandstone-dominated rock assemblages. Outcrops of the rock masses were classified using the method of *Hoek and Brown, 1997* (also *Marinos and Hoek, 2001* and *Hoek and Diederichs 2006*), as these classifications allow for comparing them quantitatively with other rock types and for assessing several model parameters. Geological Strength Indexes obtained by *Ronchetti 2007* and *Ronchetti et al. 2007* range between 50 for the outcrops of more compact flysch rock masses and 10 for the weakest clayey shales. The values of the uniaxial compressive strength of the intact rock pieces, inferred from the ratings of outcrops, as suggested by *Hoek and Brown 1997*, vary between 180 MPa for thick sandstone beds and 10 MPa for soft clayshale layers (*Ronchetti 2007, Ronchetti et al. 2007*), see Table 5, page 62.

The values for the uniaxial compressive strength of the intact rock pieces of the sandstone layers appear to be quite high, when compared to laboratory values of uniaxial compressive strength reported in literature, e.g. *Klengel and Wagenbreth 1982* or *Hoek and Brown 1980* for comparable rock-types.

Generally, any classification based on the description of outcrops is subjective to a certain extent and unfortunately, no test data of these values, that could clarify this issue, are available for the rocks of the Valoria case study. However, the descriptions of the rock-types were made by the working group that at present probably has the most profound knowledge of the field situation at Valoria. In the opinion of the author of this work, who has been to the site for several times but has not done the fieldwork, the overall classification results, i.e. together with the GSI and mi-values, with the description of the Flysch types, as well as with the data reported for the shale layers give a realistic picture of the rock mass properties encountered in the field. Therefore all classification data from *Ronchetti 2007* and *Ronchetti et al. 2007* have been taken unchanged as a basis for the derivation of the material parameters in Section 3.7.3.4.

Several soil samples of the fine-grained matrix of the landslide body (earth-slide material), taken from near the ground surface and from the inclinometer boreholes at different depths, were classified and analysed in the laboratory (*Guerra 2003, Ronchetti 2007* and *Ronchetti 2009 U*); see Table 5. The average grain-size distributions of the fraction smaller than two millimetres indicate that the matrix of the earth-slide material is composed predominantly by silt (averaging around 35%) and clay (averaging around 35%). It is further characterized by a low plasticity; with an average liquid limit of 38% and an average plasticity index of 14%. The peak cohesion values determined by means of simple-shear tests conducted in the shear box vary between 0.7 kPa and 13 kPa, the peak friction angles between 22.2° and 32.8° (*Ronchetti 2009 U*).

Two samples were taken from borehole B9A (see Fig. 31 and 32, page 71) at that depth where the inclinometer indicated the location of the main shear-zone and were sheared, also in direct shear but back and forth, until minimum values of the shear strength were obtained. After rupture was observed, the velocity of shearing was reduced to 0.0037 mm/min, which corresponds to the order of magnitude of the displacement rates measured in the field during the preparatory phases (see Section 3.4., pages 63 and 64). All shear tests were performed under normal stresses similar to the range of the overburden pressures presumed to act along the different shear-zone domains along the studied slope (see Section 3.7.2., page 81). Load steps of 320, 500 and 680 kPa were chosen for the normal stress. The residual friction angles measured in these tests were 15.1 and 15.6 degrees (*Ronchetti 2009 U*).

During the year 2005, five permeability tests were conducted inside the landslide body during the perforation of boreholes, one Lefranc test (descriptions of all performed tests are given by *Ronchetti 2007*), and in the successive four piezometer boreholes, four slug-tests. The calculated hydraulic conductivities vary from 1.3E-6 m/s to 1.2E-8 m/s, in accordance with the percentage of the clay fraction and the expected degree of consolidation of the landslide materials. Two additional tests were performed for the bedrock of the type of the clayey-silty shales formations. They gave permeability values between 1.7E-7 and 1.0E-8 m/s. For the flysch-type sandstones on-site test data was not available. Based on the observations made in the field, values reported in literature for similar rock masses were used (e.g. *Lee and Farmer 1993, Fetter 1994,* and *Gattinoni et al. 2005*) which suggest a range between 5E-5 and 1E-7 m/s (*Ronchetti 2007, Ronchetti 2009 U*).

Table 5 summarises the basic geomechanical and geotechnical characteristics of the bedrock and landslide materials giving the upper and lower values margins of the determined parameters together with the number of tested samples or conducted tests (in brackets). Where enough and sufficiently precise data is available, average values (typed in bold) are also given.

Intact rock pieces (classified after Hoek & Brown 1997 and Hoek & Diederichs 2006) / **Rock mass** (classified after Marinos & Hoek 2001) / **Rock mass** (field data and field tests)

	Estimated unconfined uniaxial compressive strength σ_{ci}	Intact rock parameter m_i	Geological strength index GSI	Composition and structure Flysch type	Modulus ratio MR	Total unit weight W [kN/m³]	Coefficient of permeability K [m/s]
Flysch-type sandstones	120-180 for sandstone layers, 20-40 for shale layers	15-19 for sandstone layers, 6-8 for shale layers	30-50	B and C	300	24-26 (estimated)	5E-5 to 1E-7 (estimated)
Clayey-silty shales	10-20	5-7	10-20	G and H	300	21-23 (estimated)	1.7E-7 to 1.0E-8 (2)

Fine-grained soil matrix (laboratory test results) / **Landslide mass** (field and laboratory tests)

	Water content w [%]	Liquid limit LL [%]	Plastic limit LP [%]	Plasticity index IP [%]	Grain size distribution of fraction < 2 mm Sand [%]	Silt [%]	Clay [%]	Cohesion at peak strength C' [kPa]	Friction angle at peak strength Phi' [°]	Residual friction angle Phi res. [°]	Total unit weight W [kN/m³]	Coefficient of permeability K [m/s]
Landslide body	**15**; 8-26 (11)	**38**; 29-59 (10)	**25**; 18-38 (10)	**14**; 9-21 (10)	**31**; 3-76 (7)	**35**; 10-54 (7)	**34**; 14-56 (7)	**8.2**; 0.7-13.0 (8)	**26.6**; 22.2-32.8 (8)	15.1-15.6 (2)	**20.6**; 18.9-22.7 (11)	**1.3E-6 to 1.2E-8** (5)

Average values (bold), value ranges, and number of tests or samples (in parentheses)

Table 5: Geotechnical parameters of the bedrock and the landslide body (*Guerra 2003, Ronchetti 2007, Ronchetti et al. 2007* and *Ronchetti 2009 U*)

The laboratory values show that the material properties of the landslide body are subject to great variations, due to the complex nature of the parent bedrock materials (see Section 3.2., pages 58 to 60). For this reason, single values tend to be meaningless, especially if the samples are not representative for the slope region of interest, and value ranges and average values are needed to describe the material properties and their variability.

In the nearby Bologna Apennines, *Simoni and Berti 2007* classified more than 170 soil samples originating from rock types of identical stratigraphic context and obtained average values of Liquid Limit of 36% and Plasticity Index of 15%. Analogous to the case of Valoria, the variability was very large, e.g. Liquid Limits varied between 18% and 63%. Furthermore, these authors found average values of the residual friction angle of around 15°, based on more than 30 shear tests performed on these samples. All these values correspond very closely with those obtained for the fine-grained matrix of the landslide material at Valoria.

3.4. Climate of the site and activity phases between 2000 and 2008

As reported by *Ronchetti et al. 2007,* and *Ronchetti 2007,* the series of measurements recorded during the last 30 years at the meteorological station of Frassinoro, which is located at three kilometres distance from the landslide at an altitude of 1100 metres yielded an average annual rainfall of 1300 mm and an average annual temperature of 10°C. Rainfalls concentrate on the autumn and spring months, the most rain-laden months are October and November (around 175 mm) followed by March and April (around 125 mm). Between 2000 and 2008, the observed peak value of the daily rainfall was about 80 mm and the maximum rainfall intensity measured during short-lasting showers exceeded 50 mm/hour. Mean daily temperatures are lowest in January and February (around 2°C) and highest in June and July (around 20°C). During winter, 2 to 4 metres of cumulative snowfall are generally observed. Snowmelt takes place prevalently from March until May. An average annual evapo-transpiration of about 650 mm was estimated by *Ronchetti et al. 2007.*

Figure 29 presents the climatic conditions near the landslide area between 2000 and 2008 together with the different phases of activity of the Valoria earthflow. These phases are described more in detail in Section 3.5., starting from page 66. The time specifications refer always to the studied slope area, the head and crown zone. From there, the movements propagated downslope in the following weeks and months. The activity of the landslide was categorised with four different terms that were defined according to the classifications proposed by *Varnes 1978, UNESCO 1993* and *Cruden and Varnes 1996.* Following these classifications, in this work it is assumed that after a major active phase or reactivation phase, termed **"critical phase"**, the landslide enters a **"suspended phase"**, in which the probability of reactivation is still elevated, because the strength along the shear-zones has been reduced by the sliding processes, so that most probably the holding and driving forces are at a very unsteady equilibrium. In this study, the term "suspended phase" was adopted to characterise the time after the reactivation event of 2001 and the time after the acceleration phase of the spring months of 2005, called **"preparatory phase"**, when the landslide was near to a critical phase; just as well for the time after the reactivations of October 2005 and 2007 (classified as "critical phases"). If during an entire hydrological year, with its seasonal changes of the climatic external factors, no new reactivation is triggered, then the landslide is, from then on, classified as **"dormant"**, assuming that the balance situation reached after the event is stable enough so that the slope can resist against at least some seasonal reductions of the holding forces (decrease of effective stress by rising pore pressures) and the shear-zone material has started to consolidate again. The above-described classification rule was used here for the time after the reactivation of the year 2001 (critical phase). After the critical phase of the 2005 reactivation event, the suspended phase was repeatedly interrupted by considerable acceleration phases without a crisis event. For this reason, this study assumed that the reached equilibrium situation was very unsteady and the shear-zone material was inhibited from reconsolidating, and so, the landslide persisted in the suspended phase more than one year without returning to the dormant phase. The geomorphic evolution and the movement rates associated with the different phases, are described in the following Sections (3.5. and 3.6.).

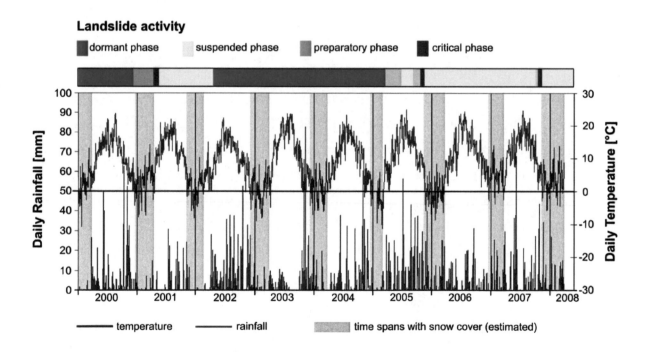

Fig. 29: Climatic conditions near the landslide site and activity phases in the head and crown zone between 2000 and 2008 (*Ronchetti 2008 U* and this study).

The synopsis of the weather conditions together with the chronology of the activity phases shows that an increase of the activity of the landslide usually takes place after the autumn months and also in the spring months. In autumn, evaporation and transpiration are decreasing due to the falling temperatures

64

and the rainfall amounts are increasing. Furthermore, the average duration of the rainfall events increases with respect to the summer. All these factors are favouring the infiltration process. In spring, the snowmelt and the high rainfall amounts, especially in March and April also lead to a rise of the pore water pressure inside the slope. Until 2008, none of the major accelerations took place during the winter time, when the slope is intermittently or permanently covered by snow (columns in light blue in Fig. 29) or the soil near the ground surface is frozen, and therefore infiltration is partially obstructed. The most important decelerations were observed during and after the summer months, when the temperatures are high, whereas the amount of precipitation is relatively low and a considerable proportion of the rainfall concentrates in short-lasting high-intensity events. As a consequence, a rise in surface-runoff and evapo-transpiration is observed, which is linked to falling pore water pressures inside the slope.

3.5. Geomorphic evolution of the slope

Development of the quaternary slope morphology

The Valoria earthflow is a prehistoric landslide. An age of 7800-7580 cal. yr BP (2σ range; *Stuiver et al. 1998*) was obtained for a wood sample from the bole of a silver fir that was collected close to the bedrock interface in the landslide toe zone (*Bertolini 2003*). Numerous radiocarbon datings of vegetation remnants indicate that landslides of the same type started developing in the Emilia Apennines at least 14000 years ago and that during the last 5000 years (Subboreal and Subatlantic time) reactivations of these were notably more frequent in certain phases which were explained by climate changes (*Bertolini and Pellegrini 2001, Bertolini 2003, Bertolini et al. 2004, Soldati et al. 2006*).

Accordingly, the geomorphic evolution of the present-day Valoria landslide most probably started in the Lateglacial (latest Pleistocene time) and the landslide was repeatedly reactivated since then. The nearby small-scale morphology was formed mainly by these reactivations of the earth-flows and the related complex phenomena, earth-slides, rock-slides, rock-block-slides, which were interacting with fluvial processes. Thereby these reactivations were of very different extension, affecting former landslide areas partially or totally (e.g. *Bertolini and Pellegrini 2001, Ronchetti et al. 2009 A*), but all in all resulted in an enlargement of the process areas that – in the opinion of the author of this work– persisted until today and will continue.

During the Cold Stages of the Pleistocene, the above-described complex landslide processes were less active. Most of the time, the study area was located very close to – but outside – the margin of the regions covered by perennial ice and snow. For example, as reported by *Guerra 2003*, at the moment of greatest extension of the Würm glaciation, the valley glacier ended nine kilometres farther upstream and the perennial snow-line was situated around 1530 mNN. For this reason, the studied slope was not preloaded or shaped by ice masses. Landforms generated during the Warm Stages, presumably resembling those of today, were smoothed during the Cold Stages by laminar and superficial periglacial denudation processes, acting only in a thin active layer. While deep-seated landslides and the processes of linear and incisive fluvial erosion declined in activity, the dominating processes – very active due to a scarce vegetation cover – were periglacial solifluction (gelifluction), favoured by the fine-grained weathering products of the marl- and claystone-formations, and periglacial sheet erosion, promoted by snowmelts and frost-shattering. Especially in those regions where the rock masses had been pre-marked by tectonic strains, freeze-thaw weathering generated many cracks and fissures (compare Section 3.2., page 59). Today, they can be observed particularly in the sandy flysch-type rocks of turbiditic origin, since their coarser and more competent layers responded to the tectonic stresses by forming distinct fracture surfaces. *Guerra 2003* reported intense cracking and fissuring of the rocks of the Arenarie di Poggio Mezzature formation, which was related to frost-shattering during the Würm glaciation. Also inside the clayey and marly formations, tectonic stresses promoted the development of discontinuities for they do not represent homogeneous beds but consist of interbedded strata of different competence (e.g. they contain intercalations of sandstones). All in all, tectonic processes scribed the working faces for weathering and therewith both for periglacial denudation processes and for the Lateglacial to Holocene slope instabilities and fluvial incision.

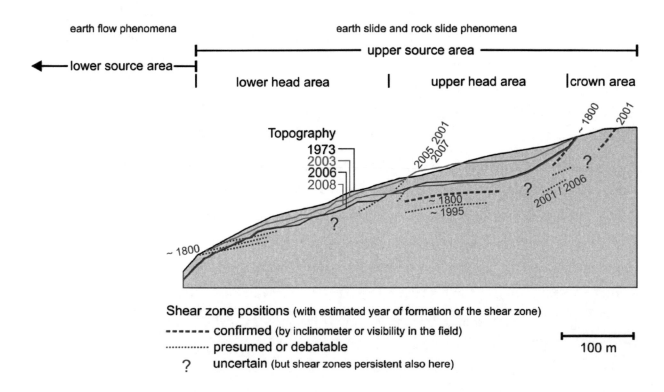

Fig. 30: Geomorphic evolution of the studied slope, i.e. the upper source area – Changes in topography and probable development of the shear-zones causing these changes (topographic data from *Ronchetti 2008 U*).

66

Information about the more recent changes of the slope morphology with time (see Fig. 30) was obtained and collected in different ways: by analysing aerial and ground-based photographs, by means of geomorphologic field surveys, and by comparing three high-resolution DEMs (Digital Elevation Models). In the following, the activity phases of the Valoria landslide between 2000 and 2008 are described according to *Ronchetti et al. 2007* and *Ronchetti 2008 U*. These phases refer to Figure 29 on page 64.

Dormant phase until the year 2000

During the last century, the Valoria earthflow reactivated several times. *Guerra 2003,* using also unpublished reports from the local geological and technical survey STBESP (Servizio Tecnico dei Bacini Enza Secchia e Panaro), analysed multi-temporal aerial photographs, which were available from the 1950s on and carried out a GIS (Geographic Information System)- based photogrammetric survey. He was able to detect the moving zones and their boundaries during the different reactivation phases and found that between 1950 and 2001, the Valoria landslide partially or totally reactivated six times (1950 and 1956 in autumn and winter; 1951, 1972, 1984 and 2001 in spring). The work of Guerra revealed that only the events starting from 2001 affected the actual upper rock-sliding and earth-sliding area, which is investigated in this study ("a" in Fig. 26, page 57, and "upper source area" in Fig. 27, page 58). Until then, the regression of the main upper scarp, mainly in the events of 1951 and 1972, had reached a maximum altitude of 1,200 metres above the sea level, where a steep step built up of flysch-type sandstones remained. This steeper section is still visible today, in the foot zone of the studied slope section (see Fig. 28, page 59).

The oldest available DEM was digitised from detailed topographic maps at a scale of 1:5000, which were made in 1973 and actualised during the 1990s. This DEM shows an intact slope profile along the studied section (see Fig. 30, page 66). Although there is some evidence in the field for possible former shallower landslide phenomena and minor rock creep, the convex profile proofs that until the year 2000 the studied slope section was not involved into the geomorphic activity of the large earth flows and the associated deep-seated complex phenomena developing in its vicinity since the Lateglacial. However, to the orographically right side of the studied profile at around 100 metres distance, a slight hollow in the slope morphology indicated settlements due to an old activation. Therefore, in the study area, small creep deformations along a zone of potential failure, connecting weaker regions of the weathered soft rocks, can be assumed for the 19th and the first half of the 20th century. The ongoing movements and also the proceeding removal of material at the foot of the slope, especially during retrogressive reactivation events led to a decrease in shear strength and finally to the progression of failure along this weak zone.

Preparatory phase 2000/2001

After abundant rainfalls in the autumn of 2000 and the spring 2001 (more than 1000 mm) accompanied by considerable snowmelt, new cracks developed in the forest below Mount Mezzature (for the location of this mountain, see Fig. 26 and 27, pages 57 and 58). They were interpreted as the regression of the landslide boundaries of 1951 and 1972.

Critical phase 2001

In April 2001, accelerated movements (up to a few metres per day) with maximum total displacements of some tens of metres, estimated from the displacements and destructions of trees and forest paths, affected the studied area. The morphology change during this event was also documented by a new Digital Elevation Model (resolution 2 m) obtained from a detailed aerial survey, performed in the year 2003 (see Fig. 30, page 66). A large amount of material was removed from the head area onto the steep step at the foot of the studied slope, where it accelerated and disintegrated. Remoulded material was transferred onto the track zone of the earthflow and the movements successively propagated until the summer, reaching the earthflow toe zone and the Dolo River (see again Fig. 27 for the locations of the different landslide-zones).

Suspended phase 2001/2002

During the dry climate of the summer season, the displacements of the whole landslide decreased. Coinciding with the autumnal wet season, a new acceleration affected the entire landslide. The deformations continued all the winter time and started to decrease in the end of spring 2002. Starting from the 2001 event, multi-temporal information about the slope morphology was collected by different authors and reported in GIS-based maps (*Guerra 2003, Manzi et al. 2004, Ronchetti 2007*).

Dormant phase 2002-2005

After the movements of the earth-flow lobes had come to a complete stop in summer 2002, only small creep deformations in the crown and head area and especially in the lower source area were observed until spring 2005.

Preparatory phase in spring 2005

In April 2005, after 576 millimetres of cumulative rainfall (since October 2004) and 296 centimetres of cumulative snowfall (between January and April 2005), extension-cracks were observed in the landslide crown area, head area and also in the lower source area. Some cracks were up to one metre wide. Simultaneously, farther downslope, earth-flow deposits were reactivated in an area of 0.35 km^2.

Suspended phase in summer 2005

From May to early August 2005, during the relatively wet summer, the cracks in the lower source area enlarged to up to two metres, while the earth-flows did not advance significantly.

Preparatory phase in autumn 2005

After 95 mm of rainfall in August, in early September 2005, the movements in the upper and lower source area accelerated and the reactivated earth-flow area enlarged to 1.0 km^2.

Critical phase 2005

Accompanied by enduring wet weather conditions (more than 150 mm of rain in September), in late September and early October 2005, a widespread reactivation of the earth flows took place, with movements propagating into older earthflow deposits, located downslope or laterally adjacent to the initially reactivated zone. A description of the further evolution of the earth flows, which nearly reached the Dolo River in December 2005, can be found in *Ronchetti et al. 2007*. In the year 2006, by means of a helicopter-based LIDAR (Light Detection and Ranging)- survey, a new DEM, with a resolution of 0.5 metres, was produced, which documents the slope morphology changes during the reactivation event of 2005 (see Fig. 30, page 66). In the surroundings of the studied section, the erosion observed in the head area adds up to tens of meters. In the field, new steps were discovered in the morphology of the crown zone.

Suspended phases and minor accelerations 2005-2007

After a relatively long period of extremely cold winter weather that started in the second half of November 2005, the soil near the ground surface had frozen or rather was covered by snow, and the movements slowed down considerably in January 2006, accelerating again in spring until decelerating anew in summer. Between autumn 2006 and autumn 2007, almost the whole landslide was no longer active, only in the crown zone and in the head zone, small deformations were observed.

Preparatory phase and critical phase 2007

Shortly after the onset of the first major autumnal rainfall period, at the end of October 2007, a new reactivation started. Thanks to the new and continuously recording monitoring instruments, for the first time also the chronology of the events of the reactivation could be reconstructed in detail. The spatial and temporal distribution of the deformations before and during the crisis phase revealed that the increased activity of the landslide developed starting nearly simultaneously from the head zone and the crown zone, followed by large displacements of the head zone, attaining tens of meters in the lower head area, moving on in the lower source area in the form of an earth flow. For the head and crown area, these findings are described more in detail in Section 3.6. (page 74), together with the monitoring data.

After the reactivation of 2007, as a new DEM was not immediately available and a further reactivation had to be expected in the near future, the topography of the studied section was surveyed in the field by means of laser distance measuring equipment and measuring tape in January 2008 (see Fig. 30, page 66). While the morphology of the crown zone had not changed significantly, erosion of up to a few metres had occurred in the upper part of the head zone whereas in the lower part, as a consequence of deposition, also originating from the lateral parts, an increase in elevation of up to 10 metres was measured.

Suspended phase 2007/2008

In the end of November and at the beginning of December 2007, the movements slowed down. In the following relatively calm phase, still considerable deformations were observed in the crown and head area (see Section 3.6., page 76).

3.6. Monitoring data of the movements and of the hydrogeological conditions

In this section, the monitoring data collected by the Applied Geology research group of the University of Modena and Reggio Emilia and by the Emilia-Romagna Region (Servizio Tecnico dei Bacini degli Affluenti del Po) between 2003 and 2008 are described according to *Ronchetti 2009 U*. Information about the rates of the slope movements was collected from the year 2003 on with different in-place instrumentations. Some of these instruments have been lost during the successive reactivation in 2005. In some cases, they have been replaced on site. Only the data obtained from monitoring instruments that were placed into the landslide head area and crown area are reported in this research, with two exceptions, namely boreholes B2A and B2B. Figures 31 and 32 display the locations of the monitoring instruments.

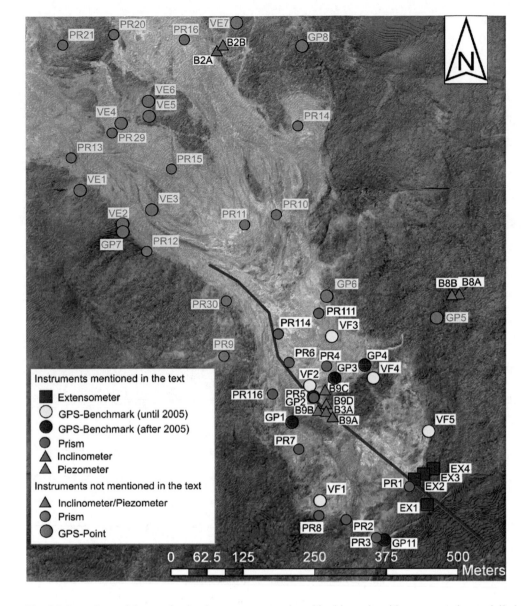

Fig. 31: Locations of the monitoring instruments mentioned in this study with respect to the modelled section (red line); graphic modified after *Ronchetti 2009 U.*

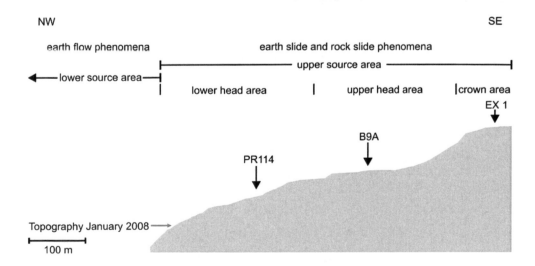

Fig. 32: Schematic actual (January 2008) slope profile along the modelled section (red line in Fig. 31). See Fig. 27, page 58, and Fig. 26, page 57, for the location of the modelled area.

Dormant phase 2002-2005

Starting from March 2003, five GPS (Global Positioning System) - benchmarks were measured with periodic RTK (Real Time Kinematics) surveys. Five surveys have been performed: in June 2003, June and August 2004, July and October 2005. Although the landslide appeared to have undergone some acceleration in the spring months of 2005, in the studied area no important deformations were detected before July 2005. In fact, GPS-point VF3 shifted only around one or two centimetres in this entire time interval. GPS point VF2, and to a lesser extent also VF4, which moved, respectively, more than 1.5 metres and about 3 decimetres before July, were not found representing the overall displacements of the head zone blocks as they were affected by shallow local ruptures observed in the field.

Preparatory phase in spring and autumn 2005

Between July and October 2005, some local phenomena showed considerable acceleration. The maximum total displacement measured was 37 metres in benchmark VF2, already marking the onset of the following critical phase. In total, the displacements measured for the years 2003 and 2004, together, were lower than those measured in 2005 until October. Summed up over this time-span, the displacements observed in the crown zone (decimetres to metres) were smaller than those of the head zone (metres to tens of meters). Again, evidence was found in the field that the latter were partly reflecting shallower phenomena. Several boreholes, thereof three in the studied slope area, were drilled in summer 2005 and equipped with two inclinometer tubes and one open-standpipe piezometer. One of the two inclinometers was placed into the central part of the active head zone (B3A), the second inclinometer (B8A) into the lateral, not active part. Near to the inclinometer borehole, an open-standpipe piezometer (B8B) was put in place. Until the first half of 2008, Inclinometer B8A had moved only a few millimetres above the depth of 12 metres. Inclinometer B3A was cut in September 2005 after one month recording time without having collected any quantitative information about the deformations. From the rupture date of the tube, it could be inferred that the displacement rate was at least 0.2 metres/month. Whereas the interpretation of the surrounding morphology, showing not even the slightest signs of new movements until this moment, suggests that this rate was not greater than 0.5 metres/month.

Critical phase 2005

Unfortunately, none of the benchmarks could be measured after October 2005 and after the 2005 event. In the studied area, the critical phase of the event lasted about 30 days until November. During this time, the entire head zone underwent movements in the range of some tens of meters.

Suspended phases and minor accelerations 2005-2007

After the 2005 event, only some of the monitoring instruments had "survived". Therefore, in summer 2007, two new inclinometers and two new piezometers were placed into the active head zone. All instruments were equipped with automatic sensors and connected by GPRS (General Packet Radio Service) to a monitoring centre. The depth of the inclinometer tubes, B9A and B9B, was 50 m and 41.5 m, respectively. The two associated piezometers were built as open-standpipe piezometers with different fissured tracts, above and below the expected depths of the shear-surfaces: piezometer B9D was fissured from 12 to 15 metres below the ground surface; piezometer B9C was perforated from 26 to 35 metres below the ground surface. Also the still existing piezometer borehole B8B, with a fissured tract between 5.5 and 24.5 metres below the ground surface, was provided with an automatic recording system and has been continuously measuring the groundwater level variations in this part of the slope since summer 2006 until today (2008). Furthermore, information about the groundwater regime in the upper part of the earth-flow track could be derived from piezometer B2B, which has been monitoring since October 2005 (see Fig. 34, page 75, for the piezometer data and Fig. 31, page 71, for the location of the boreholes). Four continuously recording wire extensometers (EX1, EX2, EX3, EX4), also equipped with GPRS mobile phones, were installed in summer 2007 inside the active crown zone, crossing open fractures.

Between autumn 2006 and October 2007, almost the whole landslide was found inactive, in the measurement period from July 2007 to October 2007, only the crown zone exhibited small movements (between some millimetres and a few centimetres). However, between 15/06/2007 and 10/07/2007, inclinometer B9A, located in the head zone of the landslide near the studied section, detected a little acceleration phase, related to rainfall events: a relative displacement of 10 mm was recorded, indicating a shear-surface at the depth of 22 metres (see Fig. 33, page 74).

Increasing deformation rates in this period of the year are not really common for this landslide type, where the movements generally coincide with high groundwater levels as a consequence of longer precipitation periods as they occur usually in the autumn and winter months and with the snowmelt in spring. An acceleration phase in the dry period of the year, when the groundwater table inside the slope is already at a relatively low level, suggests that the shear strength along the sliding-surface has reached a very low value and that already at this time, the holding forces and the driving forces along the shear-zone are at a weak equilibrium. For this reason, the little aestival acceleration phase was interpreted as a precursor to a major event in the next wet season.

During the following dry season throughout the rest of the summer and until October 2007, the installed sensors were reporting a continuous further decrease of the groundwater level and no more acceleration phases of the movements were detected. Piezometers B9C and B9D were measuring approximately constant pressure levels (-1.5 and -0.5 metres, respectively) which were generated by nearby ponds (see Fig. 34, page 75).

Inclinometer B9A

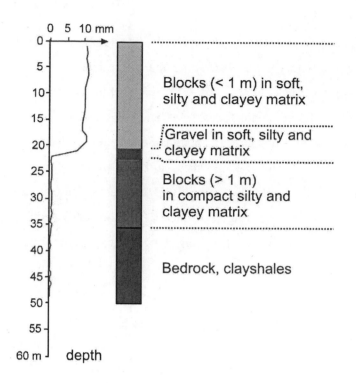

Fig. 33: Inclinometer and core profile of borehole B9A with displacements recorded during the little acceleration phase in summer 2007 (*Ronchetti 2009 U*). For the location of the borehole, see Fig. 31 and 32, page 71.

Preparatory phase and critical phase 2007

At the end of October 2007, after the first abundant autumnal rainfall period, the movements started to accelerate in the crown zone. In one week, between 24/10 and 31/10, more than 100 mm of rain with a maximum intensity of 6 mm/h had fallen (In the six months before, only 400 mm of rainfall were measured). Piezometers B8B and B2B, situated in a non active part of the landslide next to the sector that was reactivated in the subsequent event of 2007, had recorded a continuous drop of the groundwater level until then. After the above mentioned rainfall period, their pressure levels quickly increased until reaching a peak value on the 31st of October in piezometer B8B and on the 3rd of November in piezometer B2B (see Fig. 34, page 75). The continuously recording monitoring instruments allow for a closer look on the order of events during the reactivation of 2007.

One day after the initiation of the rainfall (11 mm had fallen), on 25th of October, the first acceleration of the movements was recorded by Extensometer EX2, spanned across a shallow crack in the front part of the crown zone, where a smaller block detaches from the edge of the headscarp. The onset of the acceleration of the deep-seated movements was at first detected in the central part of the head zone, where the inclinometers both started recording increased movement rates on 27th of October, first inclinometer B9A, then inclinometer B9B. Large and accelerating deformations along the shear-zone were measured from the afternoon of the 31st on. Both inclinometers broke on the 1st of November. Before being cut, the inclinometers recorded displacements of 10 centimetres along the shear-surface within one day. Until 20th of November, the upper head zone is estimated to have moved between five and ten metres, based on the morphology changes and displacements of trees and paths observed in the field.

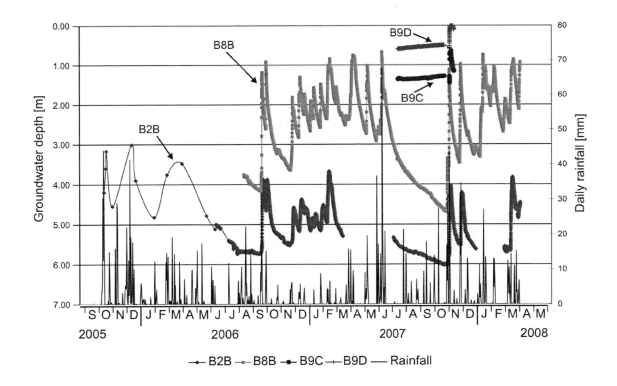

Fig. 34: Daily rainfall and piezometric curves in different boreholes during the years 2006 and 2007. Note the sudden rise after the end of October 2007, before the critical event (*Ronchetti 2009 U*).

Compared to the head zone, the increase of the displacement rate measured in the crown zone was less fast and took place more gradually. Located farther upslope from Extensometer EX2 and spanned across another fracture, Extensometer EX1 started recording accelerated movements two days later than the inclinometers, on 29th of October. Until then, 46 mm of rainfall had accumulated. Two other cracks in the crown area, monitored with extensometers (EX3 and EX4) were not affected by the event; they recorded only a few millimetres of displacements, near the range of the measurement error. Until the event, piezometers B9C and B9D, situated in the head area near the inclinometers, had not observed any fluctuations of their pressure levels, even during the preceding rainfall period, they were reporting fixed values: 1.5 metres below ground for the bedrock layer and 0.5 m below ground for the landslide layer (see also previous phase). When the inclinometers started measuring accelerated deformations along the shear-zone, the piezometric level of this part of the landslide layer suddenly started decreasing until the 31st of October (total decrease 0.15 m) , then increased by 0.5 m on the 1st of November, arriving near the ground surface. Apparently, at the beginning of the event, the zone of the piezometer boreholes was not connected to the field of rising pressure. The growing deformations first led to a sort of draining of this area, most likely by means of opening some new fissures, then, by opening further fissures or cracks, the pressure rise was able to reach also this part of the landslide body.

From October 2007 to February 2008, the displacements of 40 prisms situated on the landslide were monitored by an automatic total station, a theodolite which is situated on a stable rock ridge outside the moving slope areas. Nine periodic surveys were carried out with a measurement interval of two weeks. Two of these prisms, PR3 and PR8, had been placed into the upper part of the crown area. Between 23/10/07 and 12/12/07, that is, before and after the critical phase of the event, these prisms showed total displacements of 4 and 3 centimetres, respectively, confirming the results of the extensometer measurements: after the critical phase, the crack monitored by Extensometer EX1 had widened by 5 centimetres. In the same time span, Prisms PR1 and PR2, which had been positioned in the lower part of the crown area near the actual headscarp, had moved 34 and 32 centimetres, respectively. Again, these results are affirmed by the wire extensometers, with Extensometer EX2 reporting an opening by 12 centimetres for the detachment crack of the front block near the edge of the crown zone. Summing up the opening widths of the two cracks monitored by the extensometers and considering that there are also some smaller fractures between the main detachments which are not monitored by extensometers, because, most likely, some of these fractures are not detectable in the field, it can be concluded that both monitoring techniques have given similar results. Four prisms were located in the lower part of the head area. Due to the large slope deformations during the critical phase, two of them (PR111, PR114) were lost before having measured any displacements. Another Prism (PR116) recorded 64 metres of total displacement before being destroyed, whereas only one Prism (PR4) could still be measured on 12th of December, after the critical phase, giving a total displacement value of 23 metres.

Suspended phase 2007/2008

Only a few days after the critical event of 2007, when the movements of the head zone had slowed down, no groundwater was found in the first ten meters of the landslide body in none of the two boreholes B9C and B9D. This finding indicates that most probably the discontinuities, formed by the large deformations during the event, had augmented the permeability of the sliding body (please compare also with the monitoring data of the preparatory and critical phase). Between 20/Nov./2007 and 20/Jan./2008, the three cracks in the upper part of the crown zone monitored by Extensometers (EX1, EX3 and EX4) were all active, with a measured opening of at least one decimetre (12, 19 and 10 cm, respectively) in these two months. Directly succeeding the reactivation of October/November 2007, these measurements reflect the reaction of the crown area to the large displacements of the head area during the preceding critical phase, which had unburdened the crown. From 27th of November

2007 to 22nd of January 2008, three periodic fast static GPS-surveys were performed, separated by time intervals of two weeks. Benchmark GP11 in the upper part of the crown zone had undergone a total displacement of 33 centimetres, again in accordance with the extensometer measurements. Between 12th of December and 26th of February, Prisms PR3 and PR8, also located in this upper block of the crown zone, had been displaced by 36 cm and 6 cm, respectively, indicating a similar behaviour of this zone until the end of February. For the same time intervals, the instruments in the lower part of the crown zone, near the actual headscarp, reported the largest movements of the study area. Between 20/Nov./2007 and 20/Jan./2008, the crack of Extensometer EX2 had widened by more than seven decimetres. Interpreting this value together with the upslope cracks, a total displacement of at least one metre could be obtained for the small block at the oversteepened edge of the scarp. Also for this zone, Prisms PR1 and PR2 gave proof that the movements that had started in the preparatory phase of the critical event persisted into the following calm phase: between 12th of December and 26th of February, displacements of 58 and 78 cm were measured, respectively.

In contrary, at this time, the slowest movements of the studied area were reported by the instruments situated in the upper head zone, below the actual headscarp. The four GPS benchmarks of this area gave displacement values between 2 and 11 cm for the time interval between 27th of November 2007 and 22nd of January 2008, and were confirmed by the two neighbouring Prisms (8 cm and 12 cm between 12/Dec. and 26/Feb.). During the critical phase of 2007, the lower head area had moved faster and had received deposits from lateral parts of the bowl-shaped head zone. Afterwards, a new scarp appeared between the upper head area and the lower head area, some tens of meters farther upslope from the presumed previous boundary. In the lower head area, Prisms PR4 and PR6 gave somewhat higher displacement values (15 cm and 38 cm between 12/Dec. and 26/Feb.) than the two above-mentioned Prisms, indicating that in the calm phase the displacements of the lower part of the head area were still slightly elevated with respect to the upper part. Monitoring was continued also in the following time, which is not part of the slope reconstruction, and continues at present.

3.7. Model set up

3.7.1. Hydrogeological model assumptions

Measurements of the piezometric level near the modelled section are limited to the data from boreholes B9C and B9D and to the time span between July and November 2007 (see Fig. 34, page 75). However, Piezometers B8B and B2B, located in the neighbouring non active head zone and in the upper part of the track zone, give some quantitative information about the temporal oscillations of the groundwater level in the studied sector of the landslide since 2005 (see also Fig. 34 and Fig. 31 for the location of boreholes B9C, B9D, B8B and B2B). The lowest groundwater level measured in these boreholes was around six metres, the highest level less than one metre. The observed fluctuations accounted for one to four metres. The fluctuations can be interpreted as short-term reactions to rainfall events and to the snowmelt. During the summer, these short-term variations are oppressed by the high evapotranspiration, which constrains the infiltration process. In contrary, in winter even small rainfall or snowmelt impacts immediately lead to relevant level changes, due to the low evapo-transpiration and the high water content of the soil (near the ground surface). During the summer, the groundwater levels slowly and continuously decrease until reaching their lowest positions of the year in the early autumn.

Apart from these scarce measured data, which were gathered directly on site, the hydrogeological assumptions utilised in the models derive from groundwater-monitoring and field-test experience gained since 2001 by the Applied Geology research group of the University of Modena and Reggio Emilia in other sectors of the Valoria landslide and on other very similar Northern-Apennine landslides, which developed in a similar geological setting (e.g. *Ronchetti et al. 2006, Ronchetti 2007, Ronchetti et al. 2009 A and Ronchetti et al. 2009 B*). Comparable data were reported also by other researchers from the Apennines (e.g. *Simoni et al. 2004, Berti and Simoni 2009*) and from other parts of the world (e.g. *Schulz et al. 2009, Iverson and Major 1987, Iverson 2005*, and *van Asch et al. 2007 B*). The findings obtained on the basis of the above-mentioned test results, monitoring experiences and literature data are shortly summarised hereafter.

In the geological and climatic setting of the Northern Apennines, the groundwater level changes linked to the groundwater recharge display a seasonal trend with two peaks, the first in November-December, following a direct and lateral recharge caused by the autumnal rainfall, and the second in March-April after winterly rainfall and snowmelt. Usually, the recharge times are relatively short and vary between some days and one month. After a major recharge, the water table persists at higher levels for 3-4 months, while the regression following the discharge during the summer half year can last up to seven months. With respect to the different zones of a complex earth slide/flow, it can be stated that both groundwater depth and its fluctuation range decrease from the crown zone to the toe zone. Also the hydraulic conductivities of the materials that compose the landslide decrease in this spatial direction. As generally expected, the delay of the water level changes with respect to the rainfall input was mostly found to increase with decreasing permeability of the materials and with increasing depth of the groundwater level. In the landslide zones that are characterised by a relatively high permeability, after reaching a peak level, the pore water pressure falls again very quickly. This behaviour can be observed especially in the crown zone, where the groundwater table is generally found between 10 and 20 metres depth and shows significant variations, in the range of up to 5 metres. The monitoring experiences further suggest that in the crown zone the amount of water infiltration is controlled by fissures, cracks and pipes rather than by the porosity of the soils and rocks themselves. This finding is consistent with field-test results that revealed the presence of discontinuities attributed to lateral spreading processes where the permeability locally increases by up to two orders of magnitude

(*Ronchetti 2007*). Also *Simoni et al. 2004*, who conducted pore pressure monitoring in the headscarp and crown area of a similar landslide in the Northern Apennines (in a comparable geological setting), locally observed an anomalous quick response of pore pressures measured at depth to precipitation and mentioned the presence of open cracks or conduits as an explanation. The above-described effects are also confirmed by the observations of *Harp et al. 1990*, who investigated the variations of pore water pressure during (artificially) induced slope failures and thereby found that fractures and macropores controlled the flow of water through the slope.

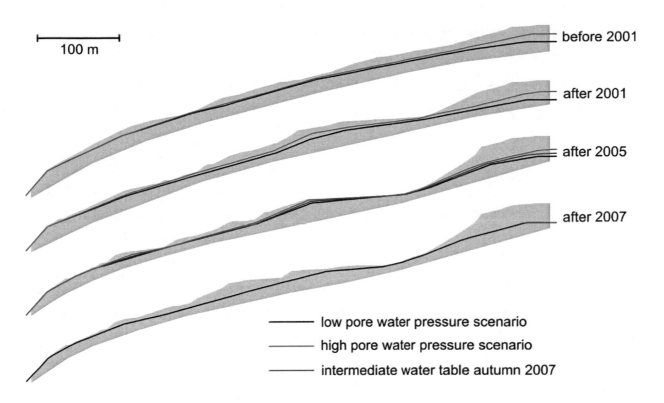

Fig. 35: Pore water pressure scenarios for wet and dry seasons and for the different slope geometries (*Ronchetti 2008 U* and this study).

In the head zone of a complex earth slide/flow, the depth of the groundwater level usually varies on a spatial basis, and is directly connected to the local morphology. In this area, the water table is typically encountered at around ten metres depth in convex slope areas and at around 2-5 metres in concave areas. In both morphological situations, the maximum changes in level range around 3-4 metres. Farther downslope, along the main landslide body and the foot zone, both the depth of the water level and the amplitude of its seasonal fluctuations normally decrease towards the toe zone, where the water table is around the ground surface and displays a low variability.

Qualitative data about the groundwater conditions along the studied portion of the Valoria landslide was collected since 2001 through continuous mapping of springs and ponds (*Ronchetti 2008 U*). By bringing it together with the data reported above, pairs of water levels were compiled for the slope geometries before and after each reactivation event (see Fig. 35). Each pair of water levels represents two scenarios, one for the dry seasons and one for the wet seasons. For the modelling of the situation in autumn 2007, also an intermediate water table was designed, because the monitoring data suggest that, at this time, most probably only a slight rise of the water pressure was sufficient to trigger a major acceleration of the movements (see Section 3.6., pages 73 and 74). After the reactivation of the year 2007, the modelling stops in January 2008, when the latest slope geometry was measured. Therefore, only one representative water table assumption was set up for this time span.

Due to the scarce measured data and the uncertainty of the information concerning the water conditions along the studied slope, the simulation of the pore water pressures is based mainly on assumptions and therefore was put into practice with the highest level of simplification possible. This means that constant phreatic levels as described in detail by Figure 35 were assigned to the different climate-driven activity phases (compare Fig. 29, page 64): the low pore pressure scenarios to the suspended and dormant phases, the high pressure scenarios to the preparatory and critical phases. Starting from these phreatic levels, the pore water pressures were assumed to increase linearly with depth.

3.7.2. Geometry Models

After the reactivation events of 2001 and 2005, two geological and geomorphological mapping campaigns covering the entire landslide area were carried out at a scale of 1:5000 (*Guerra 2003* and *Ronchetti 2007*). Additional information on the subsurface geometry was obtained by refraction seismics (*Baldi et al. 2009*) and by the examination of borehole cores (*Ronchetti 2007* and *Ronchetti 2009 U*). On this basis, two different 2D geometry models were set up along a representative section through the upper source area (see Fig. 31 and 32 on page 71) in order to account for different possible interpretations of the field and monitoring data. The first model is simplified and more robust. The second one is more complex and detailed, but implies a higher number of assumptions. Figure 36 illustrates these models with their different shear-zones and shear-zone domains. In the following, the assumptions underlying the Simple and the Complex Geometry Model are explained. Unless otherwise expressly stated, all assumptions refer to the situation at the beginning of the year 2008, which is the reference date for the reconstruction of the slope history. In Section 3.8., where the results of the modelling are presented and discussed, the two different geometry models are always placed side by side.

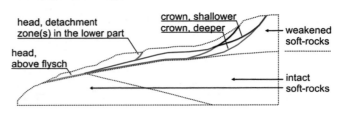

Fig. 36: Shear-zones of the Simple and the Complex Geometry Model.

3.7.2.1. Assumptions of the Simple Geometry Model

The Simple Geometry Model (see Fig. 36) is based on the following assumptions: The head zone is moving as one continuous sliding body along a shear-zone, which is located at a depth of around 20 metres and running more or less parallel to the ground surface. This shear-zone has developed long before the first major event (in the simulation the year 1800 was set) within the weakened rock mass which was covering the intact bedrock formations. In this area, the rock formations were partly disintegrated down to several tens of metres depth and their properties at that time are considered to be similar to the layer encountered in inclinometer borehole B9A between 25 and 35 metres (see Fig. 33 on page 74), which was only marginally affected by the events of 2001 and 2005. It consisted of rock blocks with diameters from one to a couple of metres, jointed by a compact clayey matrix. The largest displacements (several tens of metres) during the critical phases of the 2001, 2005 and 2007 events concentrated on this main shear-zone (shortly termed "**head**"). Its position has not changed significantly during the time span covered by the model, since due to the movements its strength underwent a further decrease with respect to the adjacent material. This hypothesis is supported by the fact that during the event of 2005, inclinometer B3A was broken approximately at the same position as inclinometer B9A during the critical phase at the end of the year 2007 (see Fig. 31, page 71, for the locations of the inclinometer boreholes).

Another pre-marked instability was presumed at the basis of the above-described disintegrated rock mass layer. Due to the big changes of the slope geometry as well as the retrogressive destabilisation of the slope after the first major event, a shear-surface started to form also here. Probably minor creep-deformations at the transition to the stable bedrock had already occurred before (in the simulation the year 1994 was used as the starting point). For this deeper shear-zone ("**head, deeper**") the simplest possible geometry was assumed, that is, parallel to the main shear-surface, at a depth where the geophysical surveys indicated the upper boundary of the stable and compact bedrock.

For the crown zone, which was unloaded and destabilised by the first major event, the Simple Model uses the simplifying hypothesis that – since the year 2001 – it separates from the slope-crest, which is still stable, along a steeply dipping detachment-zone ("**crown**"). The Simple Geometry Model represents the simplest possible structuring of the studied portion of the landslide, on the basis of the geology and geomorphology, the signs of movement found in the field, and the measurement data.

3.7.2.2. Assumptions of the Complex Geometry Model

The Complex Geometry Model (see Fig. 36) integrates additional engineering-geological assumptions, made in order to give consideration to the complexity of the phenomenon through a more detailed geometrical description. Thereby the model becomes more realistic, but engineering-geological assumptions of any kind imply uncertainties; that is why the model, by including the additional details, not necessarily becomes more correct with respect to the true situation in the field.

The geological and the geophysical investigation gave the result that the lower part of the studied slope is built up by flysch-type rocks, which are characterised by a larger portion of sandstone layers (see Fig. 28 and Section 3.2., pages 58 to 60) and that for this reason the disintegration of the rock mass could not progress so deeply in this region. As it was shown by the events of 2001, 2005 and 2007, the sliding masses could be pushed across this bar of more solid rock, but due to the different bedrock-geology the deeper – not so well developed – shear-surface is not expected to persist into this region. Therefore the Complex Geometry Model hypothesises that the shallower and the deeper shear-surface join each other upslope from this rock-bar, which means that the deeper one is only developed in the softer and more clayey rock masses (formations AUL and AVT in Fig. 28, on page 59), and at the moment (2008) the instability is advancing towards a greater depth only in this part.

Furthermore, the Complex Geometry Model comprises the assumption that, at first, the detachment of the crown zone after the sudden unloading caused by the event of 2001 occurred only via a shallower shear-zone. Before this time, the rock mass in this region, also near to the ground surface, is supposed to have been much more compact than the rock mass encountered below the main shear-zone in inclinometer borehole B9A (compare Fig. 33, page 74). The astonishing fact, that this relatively stable rock formation started moving at all, can be explained by the characteristics of this type of clay-rich interbedded strata, intensely fissured and interspersed with larger discontinuities, which – although of quite compact appearance – can hardly resist any tensile stresses. Initially these tensile stresses developed only in the shallower part of the crown zone, which was directly concerned by the unloading as a consequence of the event and the new morphology. As recently as further destabilisation and unloading were brought about by the event of 2005, the detachment zone could develop also down to the pre-marked boundary at the top of the stable bedrock and finally reach the deeper shear-surface of the head zone. Moreover, with increasing total displacements of the movements along the cracks, their width increases, opening them for water, air and weathering while, at the same time, the strength of the clay-rich material is gradually reduced by the shearing process. For this reason, in the crown zone, in addition to the deep-seated shear-zone ("**crown, deeper**"), the

Complex Geometry Model includes a shallower shear-zone ("**crown, shallower**"), which is connected to the main shear-zone of the landslide head area.

During and after the event of the year 2001, evidence was found in the field that a new shear-zone had developed inside the sliding mass while it was loosened by the deformations. Along this new shear-zone, the lower part of the head-zone started to detach from the upper part. In the shape of a step in the morphology of the slope, this detachment zone, appearing again more or less at the same position after the event of 2005, remained noticeable in the field until the event of 2007, when the lower part of the head zone was covered by several metres of deposits. Afterwards, it shifted and re-appeared slightly farther upslope. Hence, in its front part, the Complex Geometry Model contains two additional shear-zones, the first one active between 2001 and 2007, and the second one from 2007 on ("**head, detachment zones in the lower part**"). Of course, the structure of the slope is far more complicated. As an example, the monitoring data revealed that the forefront of the crown-zone, near the head-scarp, is moving significantly faster than the rest of the crown area. And along the whole slope, there are numerous other discontinuities. The Complex Model was restricted to representing only the coarsest distinguishable geomorphologic units of the studied slope.

3.7.3. Numerical Models

3.7.3.1. Applied modelling technique and discretisation of the geometry models

Numerical modelling was performed on the basis of both geometries in the framework of continuum mechanics, comprising a constitutive approach that is based on a rheological model. For the computation, the Finite Element Method was chosen and all calculations were performed by means of a well-established commercial code (PLAXIS 2D, Version 8.2, Professional, update-pack 8, build 1499). A plane-strain geometrical configuration with the real dimensions of the studied slope, 630 metres by 220 metres, was used for the analysis. Horizontal fixities were assigned to the lateral edges of the models, vertical and horizontal fixities were attributed to their basal boundaries, considering the bedrock as approximately rigid at greater depths. The different entities of the landslide were discretised as homogeneous blocks, moving along comparatively thin shear-zones. Figure 37 shows a detail of the discretisation of the shear-zones within the landslide and bedrock blocks.

The positions of these shear-zones are part of the engineering-geological assumptions of the geometry models and are therefore treated as given input of the numerical models. This approach was adopted because the available field and monitoring data include useful information on the location of these zones but provide little insight into the detailed arrangement of the different interbedded geological strata within the formations.

For building the Finite-Element mesh, triangular six-node elements, second-order elements with three stress points (Gauss-Points), were used. The blocks were meshed with the automatic procedure of the software, using a coarse setting. Along the shear-zones, the vertices of the triangular elements were locally predefined by hand in order to control the mesh size. The width of the shear-zone elements was set equal to the thickness of the modelled shear-zone layers, which was fixed uniformly for both entire geometry models at the value of 2.0 metres. The maximum length allowed for the shear-zone-elements was 10 metres. Due to the engineering-geological assumptions, the shear-zones are cropping out in the lower part of the model section. The resulting small-scale local failure mechanisms in the direct vicinity of the outcrop zones are not object of this modelling study. Therefore, the last metres of the shear-zones near their outcrops were discretised by means of interface elements.

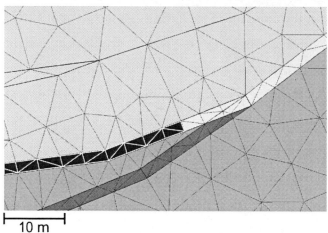

10 m

Fig. 37: Discretisation of shear-zones and landslide- and bedrock-blocks.

As a consequence of the adopted meshing practice, the numerical model resulting from the discretisation of the Simple Geometry Model, shortly named "Simple Model" consists of 1883 elements, and the "Complex Model", which comprises more shear-zones, of 2031 elements. On a standard personal computer, one calculation of the whole modelled slope history took approximately 9 minutes for the Simple Model and around 13 minutes for the Complex Model.

3.7.3.2. Material assumptions and utilised material models

The general material assumptions are based mainly on the interpretation of cores from boreholes B3A and B9A (see Fig. 31, page 71 and Fig. 33, page 74), located near the modelled section, as well as from further boreholes, which were drilled in other parts of the landslide area. Assumptions on the deformation behaviour of the materials were also made based on inclinometer measurements in borehole B9A (see again Fig. 33, page 74) and further inclinometer measurements in other parts of the slope, which exhibited a similar distribution of the deformations along depth.

The available data suggested that, except for the critical phases – when the inclinometers are broken – the largest part of the slope deformations are due to the moving of the different entities of the landslide along relatively thin shear-zone layers. These are characterised by their elevated clay and silt content. Because of the inhomogeneity of the material, the zones actually affected by the shearing process are limited to very thin domains between the harder rock fragments and their position within the weak zone varies with time. Figure 38 schematically illustrates the possible structure of such a weak zone and its interpretation in terms of the modelled blocks and layers. The material which is relevant for the deformations along theses zones, the sheared clayey soil matrix, can be described as soft and highly plastic and its behaviour is characterised by a pronounced time-dependency, i.e. it shows a significant amount of creep deformations.

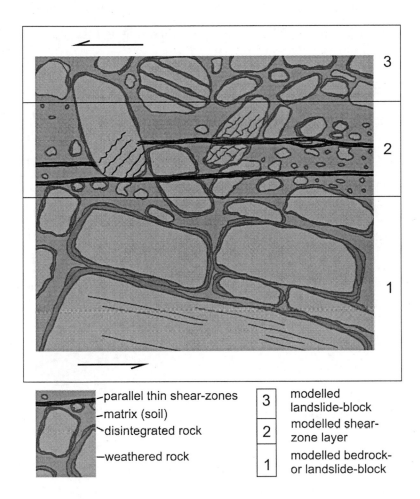

Fig. 38: Schematic picture of the geological composition of the modelled shear-zone layers.

As a constitutive model for the shear-zone material, the **Soft Soil Creep Model** (*Vermeer and Neher 1999*) was applied, because it has been developed especially for the modelling of cohesive soils and therefore takes into account some of the most important properties of these materials, namely, stress-dependency of the soil stiffness, time-dependent behaviour and failure behaviour. The Soft Soil Creep Model requires the following parameters to be specified (see Table 6).

parameter	description	(unit)
c	effective cohesion	(kPa)
φ	effective friction angle	(°)
ψ	dilatancy angle	(°)
λ*	modified compression index	(dimensionless)
κ*	modified swelling index	(dimensionless)
μ*	modified creep index	(dimensionless)

Table 6: Parameters of the Soft Soil Creep Model.

A set of three parameters (c, φ and ψ) is needed to model failure according to the Mohr-Coulomb criterion. Two further parameters are used to model the amount of plastic and elastic strains and their stress dependency. The modified compression index (λ*) represents the slope of the normal consolidation line during one-dimensional or isotropic compression. In a similar manner, the modified swelling index (κ*) is related to the unloading or swelling line. The modified creep index (μ*) serves as a measure to simulate the development of volumetric creep deformations with the logarithm of time.

The deformations of the blocks and layers between the shear-zones are comparatively small. For this reason, they are not modelled in detail. However, it is necessary to consider the load that these blocks and layers exert onto the shear-zone layers and a rough estimate of their stiffness is needed for simulating to which extent they are capable to transmit stresses and deformations. On that account, these bodies and also the stable bedrock below the shear-zones are described by a **linear-elastic material model**.

Finally, the **Mohr Coulomb Model**, an elastic-perfectly-plastic model was assigned to the interface elements at the outcrops of the shear-zones (see page 84). Shear strength and stiffness of these interface elements are assumed to be similar to those of the associated shear-zones.

3.7.3.3. Calculation phases for reproducing the slope history

In the first two calculation phases, an initial stress state is generated by applying the self-weight of the materials to the geometry model representing the oldest known topography. In this gravity loading procedure, linear elasticity is used as constitutive model and the stiffness parameters of the materials are set equal to those assumed for the recent situation. Poisson ratios are adjusted to result in a ratio between horizontal and lateral stresses similar to that corresponding approximately to a K_0-value of 0.7. After the generation of the initial stress field, the resulting unrealistic deformations are annulled and in a second phase, all materials are replaced by materials with Poisson ratios of 0.3 but with identical weight and stiffness parameters. By this means, the previously generated stress state is preserved unchanged and the following deformation analyses can be carried out with more realistic

Poisson ratios, which are expected to range around 0.3. The third calculation phase marks the starting point of the modelled slope instability. The oldest shear-zone, according to the geometry models, is inserted by replacing the corresponding clusters with material-sets accounting for soft and time-dependent material behaviour.

As the detailed loading history of the shear-zone materials is not known, the models assume materials which are in the stage of Secondary Creep and are therefore characterised by nearly constant creep rates, as long as their strength parameters are not changing. For this reason, whenever, according to the geometry models, a newly developed shear-zone is inserted into the numerical models, they are left creeping with unchanged boundary conditions during a period of Primary Creep. The durations of these periods were chosen arbitrarily in order to obtain approximately constant displacement rates with respect to the time spans modelled in the following. Thereby the estimated time intervals for which significant morphology changes could be observed in the field were used as a rough orientation: tens of years around 1800, years in the 1990s, and months/days in recent times.

After the end of the third calculation phase, the geomorphic evolution of the slope is simulated in real-time, with days as unit of time and with the year 1800 (exact numerical value: 1802) defined as the starting point. The above-described Primary Creep phases (except the third calculation phase), are part of the real-time modelling, but only the displacement rates simulated during the Secondary Creep phases are evaluated as modelling results. Modelling of the different phases of the slope evolution is performed by means of a Staged Construction procedure. This type of analysis is commonly used in Civil Engineering for the simulation of construction and excavation phases with changing load situations and groundwater conditions; also soil improvement processes (*Reference Manual PLAXIS V8 2003*). For this purpose, the model time was divided into phases, which, in a simplifying manner, stand for the phases of the geomorphic evolution observed in the field (see Figure 29, page 64 and Section 3.5., pages 65 to 70). Throughout each of the phases, the material parameters were set to be constant. As the exact course of the development of the material parameters is not known, for each considered time span a representative set of parameters was fixed based on their assumed possible temporal distribution (for details on the setting of the model parameters, see Section 3.7.3.4., starting from page 89). Water levels are changed according to the explanations in Section 3.7.1. (pages 77 to 80 and Fig. 35, page 79), always within one day and always between the different seasonal activity phases. This simple approach was adopted because the available data do not allow for determining neither the temporal nor the spatial distribution of the rises and falls and because the field measurements suggest that relatively fast changes, with respect to the duration of the phases, are possible.

The continuum mechanical method of Finite Elements (in the form as it is used here) is suitable for the modelling of displacements that are relatively small with respect to the size of the deformed elements. This applies to the dormant phases and mostly also to the suspended phases. During the preparatory phases, the modelled displacements attain the range of decimetres, whereby the calculated displacement values become imprecise. By means of an updated mesh analysis, which is based on an updated Lagrangian formulation, displacements of this magnitude can be approximated with higher accuracy (*Reference Manual PLAXIS V8 2003*). Due to the natural uncertainties underlying the boundary conditions of the problem, a fully quantitative approach is considered ineligible. Therefore, all calculations were carried out using first-order theory. The slow movement rates during the dormant, suspended and preparatory phases (maximum decimetres per month) imply a slow rate of all loading processes for these phases. Hence, fully drained behaviour is simulated, no excess pore pressures are taken into account.

Of course, the above-stated approximative simplifications do not hold true for the critical phases, when the movements along the shear-zones accelerate, causing excessive shear strains. The changes of the slope geometry and the related loading processes become faster by two orders of magnitude, producing excess pore pressures / undrained loading effects (compare Chapter 2, Section 2.5., particularly pages 49 to 52). The available shear strength is then surpassed along many discontinuities, also above the main shear-zones. The landslide blocks can therefore no longer be regarded as continua. Moreover, for the time intervals of the critical phases, when the most important changes of the slope morphology take place, the fewest measurement data are disposable, because some of the measuring instruments were destroyed by the accelerated deformations (compare again Chapter 2, Section 2.5. particularly pages 55 and 56). In conclusion, the complex processes taking place during the critical phases could not be modelled in detail. Therefore the critical phases were split up in two parts:

A first model phase simulates the situation – slope geometry, water level, and material properties – at the beginning of the critical phase. The real duration of the critical phase in the field was assigned to this model phase. The modelled displacements in this phase get very large and attain the order of magnitude of the element sizes. Usually, a converged and sufficiently accurate solution of these first calculation phases was still obtained. If this was not the case, the assumed strength parameters of the shear-zones (see Section 3.7.3.4., pages 91 to 94) were increased slightly in order to meet this criterion. (All calculations were carried out following a standard iteration procedure with specific error limits as recommended by the authors of the applied software.) The results of this first model phase allow for detecting, which blocks of the landslide models are affected by excessive deformations at the beginning of the critical phase. In these zones, and during the further course of the critical phase also further down-slope, frictional resistance forces are expected to decrease due to the large displacements leading to accelerated movements and triggering the above-described processes (e.g. undrained loading effects, additional failures within the landslide blocks). While – due to the reasons explained above – these processes are not known in sufficient detail so that they can be reproduced in the simulation, the total changes of the slope geometry that occurred during the critical phases have been precisely documented for each critical event by Digital Elevation Models and measured topographical sections.

Therefore, a second model phase simulates the total changes of the slope geometry resulting from the critical phase. There is a lack of information concerning the explicit temporal distribution of the loading and unloading which occurred during the critical events. Nevertheless, monitoring data and field observations showed that the dormant, suspended and preparatory phases are characterised by only minor alterations of the morphology, not relevant on a slope scale, and that the major changes are comparatively rapid. The maximum duration of a critical phase was 30 days while the movement rate was not constant. This means that the bulk of the changes concentrated on much shorter time intervals. For this reason, as a simplifying approach, the duration of these second parts of the critical phases was always set to one day in the models.

The total duration of the modelled "critical phases" – always including model phase one and two – was defined in such a way that it reaches until the moment when the displacement rates in the field have fallen again to low values. At the beginning of the following suspended phase, no considerable excess pore pressures are expected to remain inside the slope. Therefore both parts of the critical phases are modelled as fully drained, without taking into account excess pore pressures. Hence, the two phases just give rough pictures of what happens at the beginning and after the end of the critical phases. Anyway, the application of the new slope geometries in a staged construction process, similar to the modelling of excavation and deposition of material during constructions, allows for a quite realistic simulation of the longer-term loading or unloading of different shear-zone sections resulting from the critical events.

After the end of the critical phase of the first reactivation event, the material above the main shear-surface of the head zone is replaced by a softened material in order to account for the decrease in stiffness and strength caused by the large deformations during this phase and the subsequent critical phases (compare with the core profile of borehole B9A, drilled in 2007; in Fig. 33, page 74).

3.7.3.4. Assumptions underlying the selection of the material parameters

As it was depicted in Section 3.7.2. (pages 80 to 83), the geometry models divide the slope into different bedrock-blocks, landslide-blocks, landslide-layers and shear-zones. Table 7 summarises the derivation of the model parameters for these bodies from field and laboratory data. The underlying assumptions are explained in detail in the following.

		Total unit weight above water level	Total unit weight below water level	Stiffness	Mohr-Coulomb cohesion		Mohr-Coulomb friction angle	
					for effective normal stresses 0 - 1.1 MPa	for 0.51 MPa max. effective confining stress	for effective normal stresses 0 - 1.1 MPa	for 0.51 MPa max. effective confining stress
		[kN/m3]	[kN/m3]	[GPa]	[kPa]	[kPa]	[°]	[°]
Field data from outcrops (scale: m)	Flysch-type sandstones	24 - 26		1 - 8	220 - 600	230 - 650	50 - 62	49 - 60
	Clayey-silty shales	21 - 23		0.1 - 0.3	64 - 120	48 - 110	19 - 34	21 - 35
Modelled bedrock- and landslide-blocks (scale: 100 m)	Intact rock masses dominated by flysch-type sandstones	22	23	5	(no shear zones)			
	Intact rock masses dominated by silty clayshales			1	(no shear zones)			
	Weakened rock masses (mainly silty clayshales)	19	21	0.5	50 - 150		21 - 35	
	Disturbed rock masses (after first critical phase)	18	20	0.2	(deformations concentrating on existing shear zones)			

		Total unit weight below water level	Total unit weight above water level	Modified compression index	Modified creep index	Residual friction angle
		[kN/m3]	[kN/m3]	dimensionless	dimensionless	[°]
Laboratory data from soil samples	Clayey soil matrix of landslide body	18.9 - 22.7		0.02 - 0.04 (estimated from classification)	0.001 - 0.003 (estimated from classification)	15.1 - 15.6
Modelled shear-zone layers	Sheared soft soil	18	20	0.07	0.005	15

Table 7: Assessment of model parameters from field and laboratory data.

Weights

The material weights were assessed by means of the data presented in Section 3.3 (pages 60 to 63). Thereby the total unit weight of the rock masses was estimated from their petrographic composition and from their structure encountered at the outcrops. Greater total weights were always appointed to the materials below the groundwater table, assuming that they are fully saturated. Owing to the high clay and silt content, to the climatic conditions and to the morphologic situation in a hollow, the soils and rocks are affected by drying just temporarily and near the ground surface, while in most regions they remain nearly fully saturated due to capillary forces. Therefore the values assigned to the materials above the groundwater table are only slightly lower. For the sake of simplicity, uniform weights were given to the more intact rock masses below the landslide. The weakened rock masses at shallower depths are expected to have a higher rock-porosity and more discontinuities such as cracks that can fall dry above the water level, for this reason, they were supposed to have lower weights and the difference between the material weight above and below the water table is assumed to be larger. Again slightly lower weights were attributed to the region above the main shear-zone after the first critical event, also to the shear-zones, these materials are assumed to be in a disturbed state and therefore less compact.

Stiffness of blocks and layers

The stiffness parameters of the different bedrock- and landslide-blocks or -layers could be derived directly from the classification data presented in Section 3.3 (pages 60 to 63 and Table 5, page 62), for the concepts of *Hoek and Brown 1997* (also *Marinos and Hoek 2001* and *Hoek and Diederichs 2006*) were developed especially for estimating strength and deformation characteristics of rock masses. The values were gained at outcrops whose dimensions are in the metre-range, where typical clayshales and typical flysch-sandstones are exposed. Yet the modelled bedrock- and landslide-blocks stretch out over hundreds of metres. If representative values for these larger units are to be found, it must be considered that the formations do not represent homogeneous rock packages (see Section 3.2., pages 58 to 60), since inside the clayshales formations there are also domains of flysch-type sandstone layers and vice versa. And, for example, across the slope areas built up by the clayshales formations, besides flysch-types G and H encountered in the outcrops (see Table 5, page 62) also more compact flysch types (F, E and even D) are present in some parts. Furthermore, during the drilling of borehole cores, shales of somewhat higher quality than at the outcrops were encountered in some regions. But also in these shales, the density of fractures was very high.

For the rock masses of the flysch-type sandstones formation behind the steep step at the foot of the modelled slope (see Figure 28 on page 59), i.e. between the upper source area and the lower source area, (compare Figure 32 on page 71), stiffness parameters similar to those calculated for the materials at the outcrops were applied. Because of the prominent steep morphology, it is assumed that these sandstones are governing the material behaviour of this block.

For all the other units, which, as a result of their geologic position, are dominated by the clayshales formations, the assessed stiffness parameters lie between the bottom of the value range obtained for the flysch-type sandstones formation and the middle of that for the clayshales formations. Following the observations made in boreholes and with refraction seismics, a relatively high value was assigned to the intact bedrock formation below the landslide, and for the time after 2001, a reduced stiffness was attributed to the disturbed layer above the main shear-zone.

Starting values for the shear strength of the shear-zones

As the actual spatial distribution of the material properties along the shear-zones is unknown, each of them was divided into homogeneous sub-zones characterised by a specific set of material parameters, considering the engineering-geological setting. This division was carried out following the assumptions of the geometry models, which are described in Section 3.7.2. (pages 80 to 83), and the resulting sub-zones are given in Figure 36 (page 81). According to these assumptions, the shear-zones have formed within the weakened rock masses, consisting mainly of silty clayshales. Therefore, as starting values for the shear strength of newly developed shear-zones, typical friction angles and cohesion values as they were assessed for the silty clayshales, were set in the models. A general increase of the parameters towards those of the stronger rock masses also included in the formations was not applied, because the shear-zones are assumed to have formed within the weakest regions of the rock masses. The Mohr-Coulomb parameters were deduced again by means of the concepts of *Hoek and Brown 1997, Marinos and Hoek 2001* and *Hoek and Diederichs 2006.* In doing so, the selected stress range is of fundamental importance.

The determination of the relevant stress range was done using the numerical models of the intact slope before the development of the shear-zones. As a first approach, the effective normal stresses along cross sections, drawn through the regions where the shear-zones have formed later on, were analysed. The results suggested effective normal stresses between 0 and 1100 kPa. As a second approach, the maximum Cartesian effective stress in the out-of-plane direction in the area affected later by the shear-zones was determined, getting a value of 0.51 MPa, which was used as guiding value for the maximum effective confining stress. The resulting value ranges of the shear strength parameters for both approaches and the range of the values set in the models are presented together in Table 7 (on page 90).

At the moment when the shear-zones are introduced into the models, these zones are perceived as zones of weakness within the rock mass, whose shear strength is only slightly lower than that of their adjacencies. For setting the starting values but also the parameters at each other moment of the simulated slope evolution, the following general hypothesis was made: Inside the studied slope, strength tends to increase with depth and in upslope direction.

The main shear-zone of the head area (shortly termed "head" in Fig. 36, page 81 and Fig. 39, page 94) is assumed to have formed in the weakest regions of the rock masses. Consequently, starting values that lie near the lower boundary of the interval estimated for the silty clayshales were assigned to this zone. When following the geometry assumptions of the complex model (see Section 3.7.2., pages 82 to 83), the main shear-zone ("head") touches the flysch rock bar in the foot region, and is therefore assumed to contain more coarse components derived from the underlying sandstones and accordingly, a significantly higher friction angle is set in this part of the shear-zone (shortly termed "head, above flysch"). The effect of this higher strength is compensated by the different geometry of the complex model, where the main shear-surface ("head") is a bit steeper, converging in this part of the slope with the deeper shear-surface of the head zone ("head, deeper"). For selecting the strength parameters of the other shear-zones in the head area, medium values of the interval assessed for the rock mass strength were taken as a basis. For the shear-zones in the crown area, values near the upper margin or somewhat higher were chosen, taking into account that the destabilisation of the rock masses is proceeding in upslope direction and has not yet reached the same degree as farther downslope near the outcrops where the classifications were made.

Changes of the shear strength parameters with time

Once the failure mechanism has developed, the deformations concentrate on the existing shear-zone. Then the shear strength of the adjacent rock masses (layers 1 and 3 in Figure 38, page 85), does no longer play an important role and is therefore disregarded in the models. From that moment, the shear strength of the respective shear-zone decreases with the amount of displacements towards the residual strength of its dominating material component, that is, the clayey soil-matrix of the landslide body. The assumed decline with time of the shear strength parameters of the particular sub-zones of the shear-zones is shown in Figure 39 (page 94). For the delineation of these sub-zones, see Figure 36 (page 81) and the description of the geometry models in Section 3.7.2. (pages 80 to 83). The lines in Fig. 39 depict the assumed time course; the data points display the set values that reflect the simplified model phases, which were defined in Section 3.7.3.3. The values finally set were obtained by calibrating the models via a trial-and-error procedure on the basis of the monitoring data. Thereby the above-described starting values and the residual strength of the soil-matrix were used as absolute upper and lower limits and the engineering-geological hypotheses, which will be described in the following, were consistently observed as guiding principles.

The cardinal hypothesis demands that the shear strength along all shear-zones decreases with time and with increasing displacements, the larger the displacements the greater the decrease in strength. Given the geological composition of the modelled shear-zone layers, as it is visualised in Figure 38 (page 85), this is assumed to happen in two ways: Firstly, the alignment of platy clay particles causes a drop in strength along the shear-bands. Secondly, rotation, displacement and breaking of the coarser rock fragments effect a changing of the locations of these shear-bands into positions more favourable to sliding. A further hypothesis postulates that this decline in shear strength has accelerated considerably not earlier than in the years before the first critical event. Prior to this time, a very slow decrease of cohesion is presumed, reflecting a gradual breaking down of material bridges. This breaking down of material bridges plays an important role also in the recent times, namely in the upper part of the slope, in the uppermost head zone and in the crown zone, as well as deeper inside the slope. There, the rock masses are thought to be still in an early state of their disintegration. Temporally, the decline in shear strength is hypothesised to concentrate on the preparatory and especially the critical phases when the largest deformations take place. But this process is assumed to continue, slowly but surely, also during the suspended and dormant phases, because the slope did not stop moving in these phases.

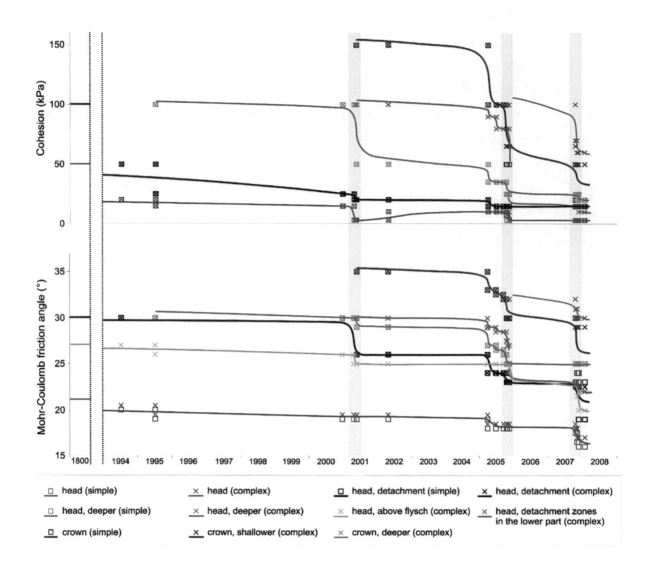

Fig. 39: Parameter assumptions for friction angle and cohesion of the shear-zones. See Fig. 36, page 81, for the positions of the shear-zone domains.

The only time span in which this principal assumption probably does not apply to all shear-zones is the dormant phase lasting from 2002 to 2005. In this time, very low movement rates were observed, perhaps temporarily the slope even came to a complete rest. According to an hypothesis of *Ronchetti 2008 U,* the clayey soil-matrix along and above the main shear-surface of the head area ("head" in Fig. 36, page 81), which was disturbed and loosened during the critical phase, is assumed to have re-compacted during this relatively long calm phase. Therefore, as unique exception to the rule of continuously deteriorating material properties, a small temporary regain of strength of this zone is taken into consideration. In the simulation, this regain of strength was expressed by a re-increase of its cohesion to the value of 10 kPa.

Dilatancy angles of the shear-zones

The material properties of the shear-zones are chiefly controlled by those of the fine-grained matrix that has formed mainly in situ between the rock blocks by the weathering of these and is characterised by high proportions of clay and silt. The original structure inherited from the degraded soft rocks has broken down due to slaking and other weathering processes, for example the oxidation of pyrite. Therefore and because of the slope history reported at the beginning of Section 3.5. (starting from page 65), this soil matrix is considered to be in a near-normally consolidated state and to show little or no dilatancy. Consequently, the dilatancy angles of the shear-zones were always set to zero.

Stiffness parameters of the shear-zones

Identical stiffness parameters were assigned to all modelled shear-zone layers, based on the simplifying approach that they are treated as if they would consist only of the clayey soil matrix of the landslide material. For the assessment of these parameters, the classification data of the samples taken from the clayey soil matrix at different depths throughout the landslide body (see Section 3.3, pages 60 to 63 and Table 5, page 62) were used as guiding values. Following the suggestions of *Vermer and Neher 1999*, it was assumed that λ^* is around 1/500 of the Plasticity Index in percent and μ^* ranges between 1/15 and 1/25 of λ^*. Values for λ^* and μ^* obtained from a series of Oedometer tests performed personally on similar materials taken from different earth-slide/flow complexes are in good agreement with the values resulting from this correlation (*Schädler et al. 2006 U*). The value of κ^* was therefore linked to the value of λ^* by multiplying this parameter by 0.5, which is the typical κ^*/λ^* ratio observed in these laboratory tests. As shown by Table 7 (page 90), somewhat higher values were used for the modelling of the shear-zone layers. As a result of the shearing processes, reworking and alignment of platy soil particles are supposed to bring about a softer behaviour of the materials along the shear-zones so that they behave less stiff than the matrix of the landslide body, at least in the direction of the shear displacements.

Model parameters for the interfaces at the outcrops of the shear-zones

The interfaces at the outcrops of the shear-zones were modelled with identical shear-strength parameters (friction angle and cohesion) as the associated shear-zones. Instead of the stiffness-parameters of the Soft Soil Creep Model, the Mohr-Coulomb Model of the interfaces requires the specification of a constant reference stiffness parameter. For these reference stiffness parameters, the default values suggested by PLAXIS as equivalent parameter sets to the stiffness parameters of the associated shear-zones were used.

3.8. Presentation and discussion of the modelling results

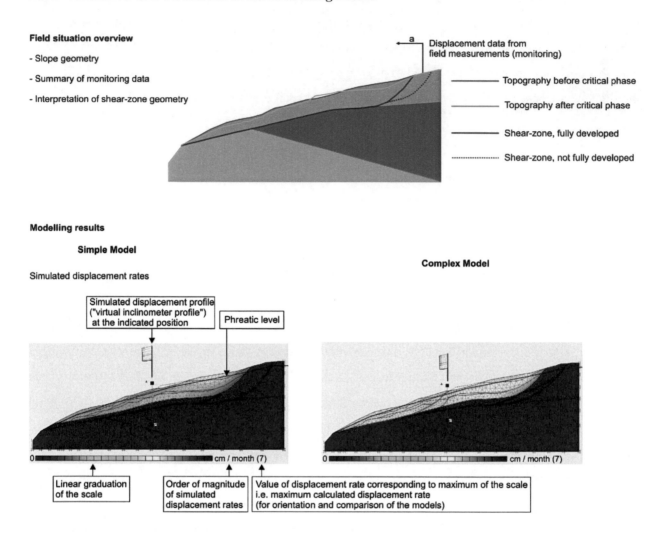

Fig. 40: Legend and explanations for the following presentations of the modelling results. For the topography before and after the critical phases, compare also Figure 30 page 66

In this Section, the results of the modelling for several phases of the slope history, selected according to their importance for the evolution of the study area and to the availability of monitoring data, are presented and discussed. First, the field and monitoring data at hand for the respective phase are presented and interpreted on the basis of an illustration. These drawings display, for each of the phases, the evolution of the slope morphology, as it was reported in Section 3.5. (pages 65 to 70) and its interpretation to depth. Furthermore they summarise the available monitoring data described in Section 3.6. (pages 70 to 77). Subsequently, also by means of graphics, the respective results of the simulations are compared to these field and monitoring data and to their interpretation, and are then discussed in the context of the model assumptions. Thereby, the left graphic shows the distribution of

96

displacement rates for the geometry assumptions of the Simple Model, the right one for those of the Complex Model. Although exact numbers are not of interest in this case study (compare Section 3.7.3.3., pages 86 to 89), for better orientation and for the comparison of the results obtained by the two different models, the values of the calculated maximum displacement rates are also given in parentheses behind the units.

Dormant phase from circa 1800 to 2000

For this phase, no monitoring data are available. Therefore some assumptions had to be made based on the field observations reported in Section 3.5. (page 67): the retrogression of the scarp in historical times, although it did not pass the flysch rock bar (brown colour in Fig. 41, see also Section 3.2., pages 58 to 60, and Section 3.7.2.2., pages 82 to 83), had surely caused an unloading of the foot area of the studied slope. Meanwhile, also weathering had progressed, especially in the zones that had been uncovered by the retrogressions of the scarp. As a consequence of this process, creep deformations started developing where the more intact rock masses that are still present at greater depths (brown and reddish colour in Fig. 41) pass into the weaker zone that has been more intensely affected by weathering and disintegration (grey colour in Fig. 41). This is assumed to entail the formation of tension cracks in a particularly weak zone, the future detachment zone (compare also Section 3.2., page 59). Though this process is of continuous nature, it was modelled in stages (1-3 in Fig. 41) with material properties and boundary conditions kept constant in each phase. One basic model assumption underlying these and all following calculations is stating that, when their behaviour in the long term is regarded, the materials along the shear-zones do not have a significant tensile strength. Therefore the tension cut-off for the shear-zones was set to zero in all calculations. As the flat part of the main shear-zone ("head" in Fig. 36, page 81) is supposed to connect the weakest domains of the weathered rock masses, low values were assigned as starting values for its friction angle and cohesion (21° and 50 kPa, respectively).

The results of the simulations based on these assumptions are shown by the first pair of graphics in Fig. 41. At this stage, which represents the situation around the year 1800, the Mohr-Coulomb criterion is reached only in a few points, but in the detachment area, numerous tension cut-off points indicate the formation of tension cracks. The modelled displacement rates are in the range of millimetres per year for both geometry models.

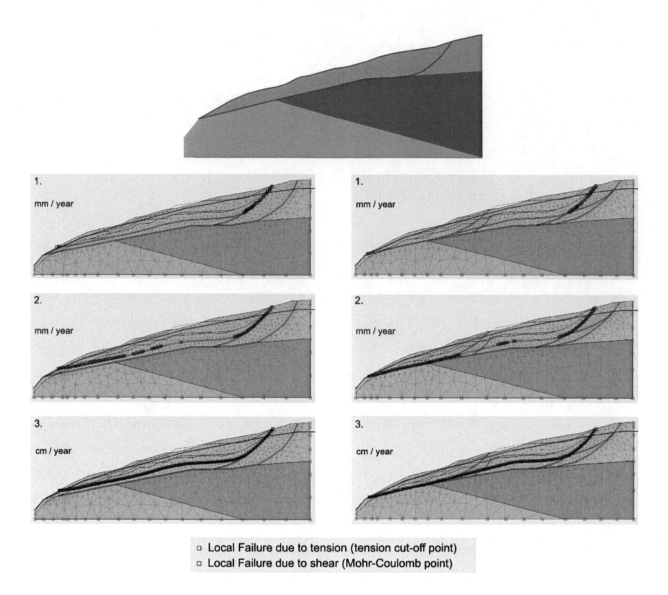

Fig. 41: Field situation and simulation results for the dormant phase between 1800 and 2000 (topography data from *Ronchetti 2009 U*). Results are presented for 3 time spans: 1) 1800-1940, 2) 1940-1990, and 3) 1990-2000.

It was assumed that movements with similar displacement rates have taken place also in the field and lasted over more than 100 years. And that, as a result, several material bridges were destroyed, making possible a straighter course of the shear-zone through the rock blocks. Based on these assumptions, after 140 years, in a second stage of the modelling (phase 2 in Fig 41), the cohesion of both the flat part ("head" in Fig. 36, page 81) and the detachment part ("head, detachment" in Fig. 36, page 81) was divided approximately in half (to 20 kPa and 50 kPa, respectively), and the friction angle of the flat part ("head") was reduced by one degree (to 20°).

In the simulations, this leads to an increase of the creep rates to roughly the double, though the calculated displacements still are in the range of millimetres per year. As shown by the second pair of graphics in Fig. 41, failure due to tension is still happening in the upper part of the detachment zone. In addition, the Mohr-Coulomb criterion is reached in large areas of the shear-surface.

The models therefore presume that in the next 50 years, until the 1990s, with the movements, also the loss of strength with time had continued and accelerated, destroying still more material bridges in the detachment zone. Along the flat part of the shear-zone, it is assumed that these material bridges were already largely lost at this time, and finally the shear-zone predominantly went through materials that were similar to the actual soil-matrix of the landslide-body. Hence, for the third phase of the modelling (stage 3 in Fig. 41), the cohesion in the detachment zone ("head, detachment" in Fig. 36, page 81) was again divided in half (to 25 kPa), and in the flat part of the shear-zone ("head"), cohesion and friction angle were further reduced. The values set for this zone (19° for the friction angle and 15 kPa for the cohesion) are based on the assumption that at this time, the friction angle of the main shear-surface of the head area was only a few degrees above that characterising residual strength of the soil-matrix of the landslide body.

The simulation results under these assumptions are presented by the third pair of graphics in Fig. 41. At this stage, the shear strength described by the Mohr-Coulomb parameters is exceeded in a lot of points along the entire length of the sliding-surface. The calculated displacement rates are higher by one order of magnitude, now in the range of centimetres per year for the low pore water pressure scenario (see Fig. 35, page 79). The simulation results for this phase suggest that in this situation, larger accelerations of the movements are imminent, because significantly higher displacement rates are to be expected during the repeated seasonal rises of the water level and the growing total displacements will effectuate a further decline in shear strength.

Preparatory and Critical phase 2000 and 2001

Also for these phases, no measured data are available, just cracks were observed in the field and the morphology changes during the critical phase are documented by DEMs. Considering the meteorological situation (compare with Section 3.5., page 68), it can be inferred that the water level was very high, probably somewhat higher than the typical water level for the wet seasons.

In the simulation of the preparatory phase, the pore water pressure was raised to this level (see Fig. 35, page 79). Without any change of the rest of the parameters and boundary conditions, this led to a drastic increase of the displacement rates. Along the main shear-zone of the head area, they rose by a factor of 30 to 40, to the range of centimetres per month. Along the deeper shear-zone ("head, deeper" in Fig. 36, page 81), nothing happened (see Fig. 42).

Regarding the initiation of the critical phase, it was assumed that beforehand, high pore water pressures prevailed over several months (see Section 3.5., page 68). In the models, this assumption causes total displacements in the range of decimetres. In the field, visible cracks appeared in the upper head zone after several very wet months. After these displacements, the shear strength is expected to have declined again, with practically no cohesion being available anymore along the shear-zone of the head area. At the most, the developing sliding body could still have got caught to a minor degree in the upper part, where the rock mass is harder. In the simulation, the cohesion along the flat part of the shear-zone ("head" in Fig. 36, page 81) was reduced to approximately zero and a low cohesion value (20 kPa) was assigned to the zone of detachment in the upper part ("head, detachment" in Fig. 36, page 81).

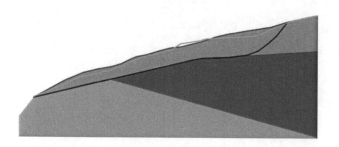

Preparatory phase 2000 / 2001

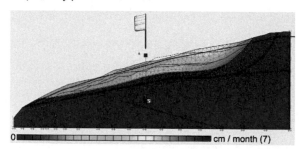 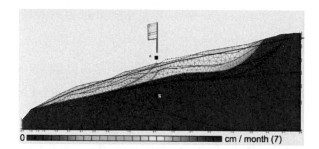

Critical phase 2001

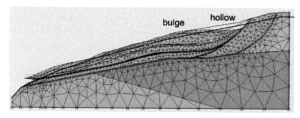 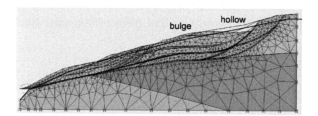

Fig. 42: Field situation and simulation results for the preparatory phase and for the beginning of the critical phase in 2000 and 2001 (topography data from *Ronchetti 2009 U*).

In the models, these assumptions result in very large displacements, that cannot be calculated precisely anymore by the approximation procedure (for details on this, see Section 3.7.3.3., pages 87 to 88). For both models, the obtained displacement rates are in the range of metres per month, suggesting that they have increased again by two orders of magnitude. This situation can be interpreted as the beginning of the critical event. The qualitative distribution of the displacement vectors at this time is presented in Figure 42 with exaggerated displacements in the form of the deformed mesh. Analogous to the new DEM measured after the critical phase (the line in light green in Fig. 42), these graphics show a hollow near the detachment and a bulge a bit further downslope. Thus, the models provide a plausible explanation to the astonishing fact that, despite the large erosion and overall loss of material, the regions slightly below the new scarp appeared as elevated terrains after the event.

Dormant phase 2002-2005

The data detailed in Section 3.6. (page 72) are interpreted in the way that throughout the entire dormant phase between 2002 and 2005, the deep-seated movements (affecting the slope to a depth of more than 10 metres) add up to only some centimetres. This would mean that the landslide has practically come to a standstill, except for minor creep deformations, that were affecting the weathered and loosened zone above the intact bedrock (compare borehole B9A in Fig. 33, page 74, above the depth of 35 metres). The much larger values measured by VF2 and VF4 were ascribed to local instabilities in the rough post-event morphology, resulting in shallow ruptures or slides (characterised by a depth of some decimetres or maximum 2-4 metres).

It is probable that during the critical phase of 2001, to a little extent, also the layer between the main shear-zone and the stable bedrock was drawn along with the bigger sliding body above it. For this reason, it is assumed that also the shear strength of the deeper shear-zone of the head area ("head, deeper" in Fig. 36, page 81) has decreased in consequence of the event of 2001. Therefore in the simulation, the cohesion value for this zone was divided in half (to 50 kPa). Certainly the event of 2001 had brought about great unloading in the head zone and had caused a loosening of the rock mass, which was accompanied by the formation of new fissures and cracks. In the following years, water and air and consequently weathering were therefore able to penetrate farther and to affect also the layer below the main shear-zone to an increasing degree. According to the hypothesis of this work, the groundwater level in the head area has fallen considerably due to the newly-formed discontinuities. In connection with the changed geometry, the upper head zone now being lower and the head zone all in all less steep, this led to a pronounced deceleration of the movements.

In the simulation of the suspended phase (no illustration), after the drop of the water level to the low pore water pressure scenario following the event of 2001 (see Fig. 35, page 79), displacement rates in the range of centimetres per year for the shallower and millimetres per year for the deeper sliding body were calculated by both models.

In the field, during the late dormant phase, quite adverse water conditions were observed in the study area during the monitoring of springs and ponds, and it seems that temporarily they were even worse than at the moment of the triggering of the reactivation event of 2005. Surprisingly, at this time, no memorable accelerations of the movements were detected. Far from it, until spring 2005, the displacements were smaller than in the suspended phase between 2001 and 2002, when the water table was much lower. Also in the models, significantly elevated displacement rates are calculated for this type of scenario (no illustration). It is therefore assumed that during the very calm phase after the event of 2001, the relative movements along the main shear-zone of the head area ("head", in Fig. 36, page 81) virtually came to a rest. As a consequence, the material along and above this shear-zone had time to reconsolidate and to become more compact again, regaining a very little but noticeable part of its strength.

In the simulations presented in Figure 43, the cohesion of the flat part of the main shear-zone of the head area ("head" in Fig. 36, page 81) was re-increased to a low value (10 kPa). This led to a substantial overall decrease of the displacement rates with respect to the suspended phase: from centimetres per year to millimetres per year, for the low pore-water-pressure scenario (see Fig. 35, page 79). As demonstrated by the simulated displacement profiles (in Fig. 43), the relative movements along the aforementioned shear-zone nearly stopped. The GPS measurements (see "a" in Fig. 43) also exhibited small movements of the crown zone. During the event of 2001, a new morphologic situation has occurred there, and the edge of the crown was unloaded by erosion.

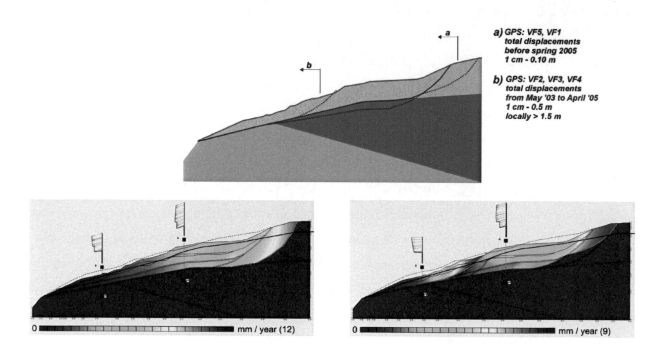

Fig. 43: Field situation and simulation results for the dormant phase between 2002 and 2005.
(Field data from *Ronchetti 2009 U*).

This explanation is supported by the models. In the calculations shown by Figure 43, high values for cohesion and friction angle (150 kPa and 35°, respectively), as they are attributed to the undisturbed rock mass (see Section 3.7.3.4., page 90), were applied along the shear-zone of the crown area. Nevertheless, both models display slow creep deformations in the crown area, outputting rates of around one centimetre per year. The complex model additionally features elevated movement rates below the secondary detachment zone between the lower head area and the upper head area that had formed during or after the critical event of the year 2001.

In this context, the alternative hypothesis that this shear-zone could have existed already before 2001 was also tested. As for this time the field data at hand is sparse, this is a thinkable assumption. However, in this case, the history of the slope can not be reproduced correctly in the simulation. Even when unrealistically high values for the shear strength of this zone are entered, extremely large displacements are calculated already for the preparatory phase in the year 2000 and, in the first instance, only the front block is affected. In contrast, the field observations made before and during the event of 2001 indicated quite clearly that from the beginning also the upper head area was just as active as the lower head area. It is rather presumed that the landslide body in the central part of the head area was loosened as recently as by the large deformations in 2001 and that before, it was much more compact and its shear strength was higher than that along the main detachment zone ("head" in Fig. 36, page 81). For the upper detachment zone had already been weakened by its tectonic history and the long-lasting movements.

Preparatory phases in spring and autumn 2005

The monitoring data specified in Section 3.6. (page 72) were interpreted as follows: The displacements were largest in the lower head zone and smallest in the crown zone. The first acceleration took place in spring due to the elevated pore water pressure. In summer, the seasonal decrease of the pressure level caused a slow-down of the movements. Already at the beginning of September, the water level was high again and the movements accelerated. The cumulated displacements of the year 2005 implicated a proceeding decline of the shear strength along the shear-zones, particularly in the crown area, which until then had moved only some centimetres. At this time, the foot of the crown zone was more and more unloaded by the movements of the head zone. On that account, the displacements started increasing also in the crown area.

In the simulation, after the rise of the water level to the scenario attributed to spring and autumn of 2005 (see Fig. 35, page 79) and after a reduction of the shear strength parameters, the displacement rates grew significantly. In the head zone, the calculated rates are exceeding those of the simulated dormant phase by one order of magnitude in the case of the Simple Model and by two orders of magnitude in the case of the Complex Model. The displacements computed for the crown zone are similar in both models, one to two centimetres, when added up over the three months of the model phase. The assumption of an additional detachment zone in the front part involves higher displacement rates in the head area than the assumption of the geometric circumstances of the Simple Model, where the holding forces in the flatter upper head area have more effect on the whole continuous landslide block.

If the water level is lowered to the low pore-pressure scenario of the summer, even if simultaneously the shear strength parameters are further reduced, the modelled displacement rates fall remarkably. For this constellation, both models output movement rates of a few centimetres per year (no illustration) and in this way they consistently explain the typical summerly calm period, also for the year 2005. A re-increase of the water level to the high pore pressure scenario (see Fig. 35, page 79) brings about movement rates similar to those obtained for the preparatory phase in spring 2005 (no illustration).

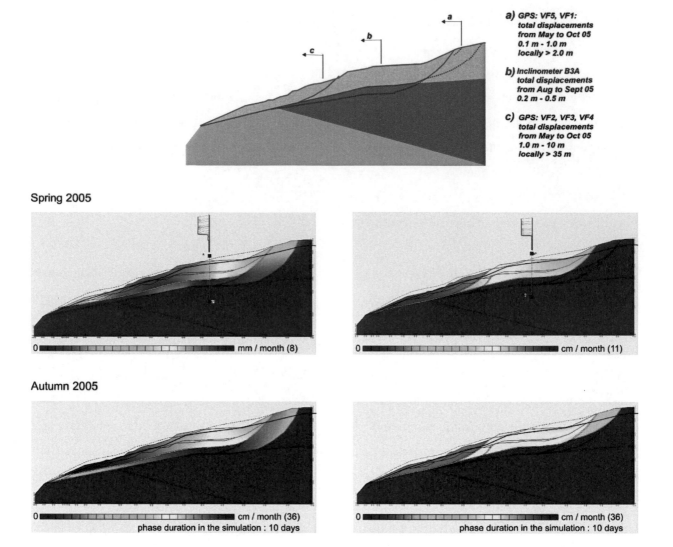

a) GPS: VF5, VF1:
total displacements
from May to Oct 05
0.1 m - 1.0 m
locally > 2.0 m

b) Inclinometer B3A
total displacements
from Aug to Sept 05
0.2 m - 0.5 m

c) GPS: VF2, VF3, VF4
total displacements
from May to Oct 05
1.0 m - 10 m
locally > 35 m

Spring 2005

mm / month (8)

cm / month (11)

Autumn 2005

cm / month (36)
phase duration in the simulation : 10 days

cm / month (36)
phase duration in the simulation : 10 days

Fig. 44: Field situation and simulation results for the preparatory phases in 2005 (field data from *Ronchetti 2009 U*).

By adding the assumption that, during the spring and summer, shear strength, mainly in the crown zone, had again decreased, displacement rates in the range of decimetres per month were achieved in the simulation of the preparatory phase of autumn 2005. Thus, at least qualitatively, the models are able to explain the drastic increase of the displacement rates in the field in autumn 2005, which also had caused the destruction of inclinometer borehole B3A (see Fig. 31, page 71, for the location of this borehole). In the calculations presented by Figure 44, when compared to the strength ascribed to the intact rock mass, the cohesion value for the shear-zone of the crown area has been reduced to one third, and the friction angle has been lowered by five degrees (to 50 kPa, Complex Model: 65 kPa, and 30°, respectively). The results of the simulations for this phase are very similar for both models, because at this time the deformation rates in the head zone depend to a lesser extent on the geometry of this zone but mainly on the forces that the unstable crown block is exerting on it.

Critical phase in autumn 2005

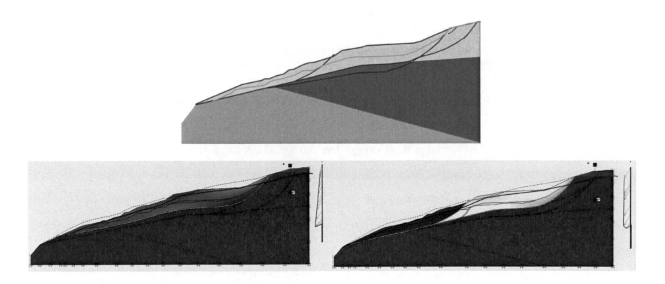

Fig. 45: Field situation and simulation results for the beginning of the critical phase in 2005.
(field data from *Ronchetti 2009 U*).

The principal cause for the initiation of the critical phase in autumn 2005 is seen in the increasing deformations during the preparatory phase, which have degraded the shear strength along the shear-zones while pore water pressures remained high. The comparison of the DEMs of 2003 and 2006 (green lines in Fig. 45) revealed that during this phase the entire head area has undergone great erosion, loosing up to more than 20 metres in altitude. From this erosion, it can be concluded that during the critical phase, the head area has moved several tens of metres and that these movements were not limited to the modelled shear-zones but, in addition, large internal deformations have affected the landslide mass in this area.

In the simulation, after having reached the situation of the late preparatory phase illustrated by Figure 44, a very little diminution of the strength of the main shear-zone of the head area ("head") is sufficient in both models to effectuate extremely large displacement rates. Figure 45 shows the qualitative distribution of these displacement rates for the assumption that the friction angle of the main shear-zone ("head") was one degree lower than at the beginning of the event of 2001 (also with negligible cohesion).

As long as the shear strength parameters assigned to the crown zone are not reduced substantially, these excessive displacements are confined to the head zone in both models. Qualitatively, this finding is in accordance with the real morphology changes observed in the field (see the upper part of Fig. 45). In the opinion of the author, lower shear strength parameters in the crown zone – lower than those assigned for the late preparatory phase – are unrealistic because this formerly intact rock mass region was concerned by the deformations of the landslide not before the critical event of 2001 (compare also the remarks to the previous phase, page 103). Also for the beginning of the critical phase, the Complex

Model with its additional shear-zone in the front part displays larger movement rates for the lower head zone than for the upper head zone.

Suspended phases and minor accelerations 2005-2007

The monitoring data described in Section 3.6. (pages 73 to 74) and summarised by the upper part of Figure 46 are interpreted in the following way. The measurement period between July 2007 and October 2007 is part of a calm phase with minor seasonal accelerations that lasted for two years. From July to October, the displacement rates of the crown zone block as well as those of the upper head zone were in the range of centimetres per year. Extensometer EX2 (shown by "b" in Fig. 46), represents a shallower local mechanism, which is not subject of the models, namely the separating of the edge of the steep scarp that was laid bare during the critical event of 2005 from the rest of the crown zone (compare Section 3.6., page 74).

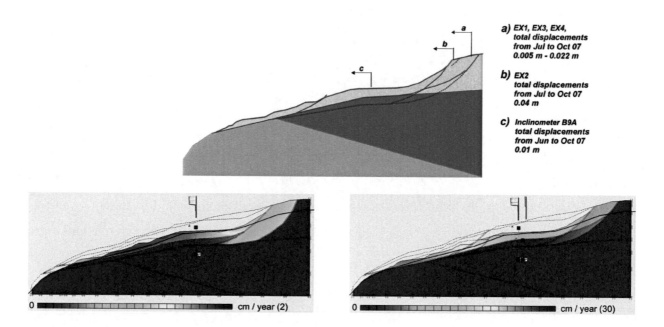

Fig. 46: Field situation and simulation results for the suspended phases between 2005 and 2007. (Field data from *Ronchetti 2009 U*).

All in all, the low movement rates from 2005 to 2007 are attributed to a drop of the pore water pressure to a considerably lower level after the critical event of 2005. There are two reasons attributed to that drop, which are both connected with this preceding critical event. First of all, many new fissures and cracks have been generated by the large deformations, intensifying the draining of the head area. Secondly, the head zone was deepened, causing an increased drainage of the crown area, and probably also the erosion in the foot area had a smaller but similar effect on the head area. Furthermore, independent of the event, in the measurement period that is considered here, also the seasonal trend with its lowering of the water level in summer plays an important role.

106

The aforesaid overall hypothesis is supported by the models: If the water level is dropped to the low pore pressure scenario (see Fig. 35, page 79), the displacement rates calculated for the upper head zone and for the crown zone are only in the range of centimetres per year, in both models. This still holds true if the shear strength parameters along the main shear-zone in the lower head area ("head", "head, above flysch" in Fig. 36, page 81) are slightly reduced, as it was done in the simulations presented by Figure 46. This reduction is based on the assumption that, as a consequence of the large material transport during the critical phase of 2005, a larger portion of weaker materials, originating from the weathering of the clayshales formations farther upslope, were occupying this area from that time on.

The calculation with the Complex Model (right side in Fig. 46) includes the additional assumption that also after the critical event of 2005 there was a detachment zone separating the upper and the lower head zone. Taking into account the large internal deformations of the landslide body during the critical phase of 2005, very low values, more characteristic for a clayey soil than for a rock mass, were used for friction angle and cohesion of this zone (23° and 15 kPa, respectively). Analogous to the modelling results obtained for spring 2005 (compare Fig. 44, page 104, with Fig. 46), the geometry of the Complex Model results in a higher mobility of the landslide blocks, that is why the Complex Model displays significantly higher displacement rates for the upper head and crown zone (5 to 15 cm/year) than the Simple Model (1 to 2 cm/year).

Qualitatively, the displacement profiles in Figure 46 are in agreement with the measurements of inclinometer B9A, which was located approximately at the same position next to the profile: Above the shallower shear-zone of the head area, situated at a depth of circa 20 metres, the simulated displacement rates are in the range of centimetres per year and constant with depth. Below this zone, both models output rates of only a few millimetres per year, which are smaller by more than one order of magnitude in each case.

Preparatory phase and critical phase 2007

As described in Section 3.6. (page 74), the availability of continuously recording instruments, allows for a less tentative interpretation of the data for this time span, which will be given in the following. The preparatory phase lasted only a few days. At this time, accelerated movements started more or less simultaneously in the head zone and in the crown zone. After seven to ten days (depending on the way the phases are defined), the critical phase started, involving displacements in the range of metres in the head zone, which had been weakened, on an on, by the preceding events. Furthermore, over the course of the suspended phases from 2005 to 2007, repeated seasonal acceleration phases had been observed. Thereby the crown zone had been destabilised more and more. Consequently, the seasonal water table rise in autumn instantaneously triggered accelerated movements, which resulted in a further decrease of the shear strength and thus in a further acceleration that finally led to the initiation of the critical phase.

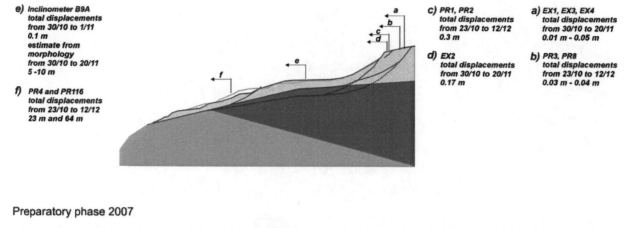

Preparatory phase 2007

Critical phase 2007

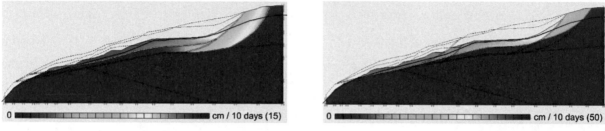

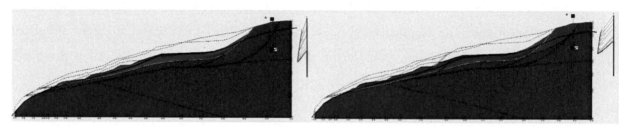

Fig. 47: Field situation and simulation results for the preparatory phase and for the beginning of the critical phase in 2007. (Field data from *Ronchetti 2009 U*).

In the simulations presented by Figure 47, according to the above-described interpretation of the monitoring data, the strength parameters of all shear-zones were reduced with respect to the assumptive situation immediately after the critical event of 2005. The parameters of the shear-zones in the head area were reduced slightly, those of the shear-zones in the crown area drastically. This was done to such an extent that already the slight rise of the pore water pressure from the low scenario to the intermediate level in Figure 35 (page 79) was sufficient to cause displacement rates that exceed those of the simulated suspended phase by two orders of magnitude.

Due to the differing number and geometry of the shear-zones (see Section 3.7.2., pages 81 to 83), with the Complex Model having two shear-zones in the crown area at this time, a much smaller decline of the shear strength in the crown zone is needed in this model to induce this great increase of the displacement rates and thus to reproduce qualitatively the incidents in the field.

Following the interpretation of the monitoring data, for the simulation of the beginning of the critical phase, the shear strength parameters were further reduced slightly and the water level was risen to the high pore pressure scenario (see Fig. 35, page 79).

Thereafter, both models displayed excessive displacements in the head area. In the case of the Complex Model, unlike the Simple Model, this reaction is obtained also without the further rise of the water level, just by a very small reduction of the shear strength parameters. This is, again, because the geometry assumptions of the Complex Model, with the additional shear-zones, provide more mobility to the blocks and layers. Therefore the simulation with the Complex Geometry Model suggests that even under more favourable weather conditions, with a smaller rise of the pore pressures, the critical event would have taken place in autumn 2007. In the light of the field observations made in summer 2007 (see Section 3.6., page 73), this proposition appears to be not so far from reality.

Suspended phase 2007/2008

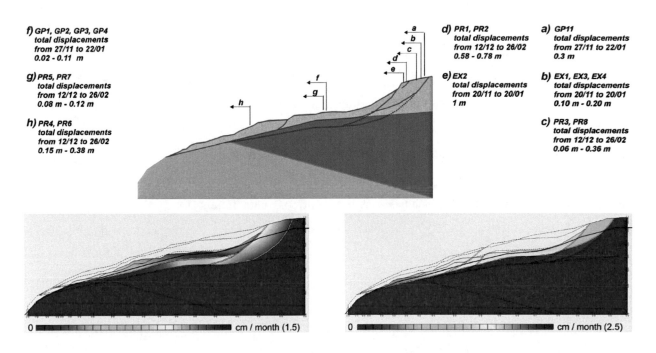

Fig. 48: Field situation and simulation results for the suspended phase 2007/2008 (field data from *Ronchetti 2009 U*).

The monitoring data detailed in Section 3.6. (pages 76 and 77) and shortly presented in Figure 48 are interpreted in the way that after the critical event of 2007, the crown and head area still showed considerable displacement rates, in the range of several centimetres per month. In the front part of the crown zone, which is not subject of the reconstruction (see "d)" and "e)" in Fig. 48), even larger displacements have occurred in the measurement period. The movements in the head zone have slowed down after the critical event, because the water level has fallen significantly as a consequence of the new fissures and cracks, which had formed due to the large deformations (for the background of this interpretation, see the description of this phase in Section 3.6., page 76). The crown area, where shear strength has decreased a lot before and during the critical phase, had been further unloaded by the large displacements of the head area. For this reason, after the critical phase, the movements in the crown area continued with similar displacement rates also in the suspended phase.

In the simulations, if the water table is lowered to the low pore pressure scenario that is attributed to the time after the event of 2007, (see Fig. 35., page 79) both models calculate displacement rates in the range of not more than centimetres per month, for the head area as well as for the crown area. This is still the case, if the shear strength of all shear-zones is again slightly reduced, as it was done in the calculations presented in Figure 48, assuming that the large deformations during the critical event have diminished the strength of the shear-zones. Hence, the model hypotheses are able to explain qualitatively also the fall of the displacement rates after the critical event of 2007 and the finding that in the subsequent suspended phase, these rates were of the same order of magnitude in the crown area as in the head area.

Comparison of the results obtained with the Simple Model with those of the Complex Model

Both models are able to reproduce the slope history qualitatively and give quite similar results, but also some differences can be noticed: For the time until the year 2001, results of the Simple Model and the Complex Model are approximately the same. For the dormant phase between 2002 and 2005, the Complex Model shows larger displacements in the lower head area. For the acceleration phase of spring 2005, the Complex model outputs larger displacement rates for the entire head area. For summer and autumn 2005, the results of both models are similar, except for the critical phase of 2005, when considerable relative displacements at the additional detachment zone of the Complex Model can be observed. Between 2005 and 2007, the Complex Model simulates considerably larger displacement rates than the Simple Model. In the preparatory phase of 2007, the decline in shear-strength (in the crown zone) that causes a comparable increase in displacement rates in both models is smaller in the case of the Complex Model; and in contrast to the Simple Model, excessive displacement rates at the beginning of the critical phase of 2007 are reached without the further rise of the water level. At the beginning of 2008, results of the two models are again similar, but displacements in the lower head area are larger in the Complex Model.

However, when considering the uncertainties that are related to the model assumptions, the results obtained by the two models can be regarded as equivalent. Whether the Simple Model or the Complex Model is more realistic, depends to a large extent on the interpretation of the monitoring data (to which the modelling results are compared). For example, between 2005 and 2007, due to the destruction of devices during the preceding critical phase, measurement data is limited to the time span between June and October 2007, and it is assumed (but not known) that displacement rates were similar during the rest of this time interval. In the opinion of the author and of one of the geologists who are carrying out the monitoring campaign in practice (*Ronchetti, F., University of Modena and Reggio Emilia, personal communication*), the results of the Complex Model are more realistic.

Analysis of the situation in 2008 and consequences for the following years

Figure 49 illustrates the field situation at the beginning of 2008 and the expected consequences for the following years, which will be described below. The movement rates that were still measured at that time in the crown and head zone, in spite of the relatively low water table (compare section 3.6., page 76), were remarkable. They indicated that the shear strength of the rock masses along the shear-zones had reached a very low level, in the head zone because of the repeated large deformations during the critical events of 2001, 2005 and 2007, in the crown zone due to creep that was caused by the consequential gradual unloading.

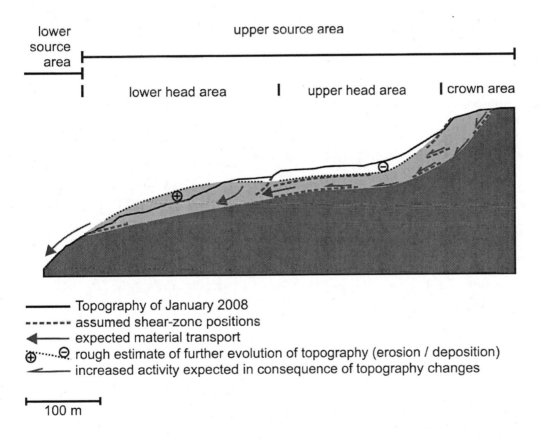

—— Topography of January 2008
------- assumed shear-zone positions
◄—— expected material transport
⊕·······⊖ rough estimate of further evolution of topography (erosion / deposition)
—◄— increased activity expected in consequence of topography changes

├———┤
100 m

Fig. 49: Field situation at the beginning of 2008 and expected consequences for the future.

The model parameters that were set at the end of the simulated time span are based on the assumption that, at that time, the shear strength along the entire main shear-zone of the head area ("head" in Fig. 36, page 81) is in the range of the residual strength of the clayey soil matrix of the landslide body (compare Table 5, page 62). This means that, onward from that time, this shear-zone went straight

111

through nearly homogeneous soil material (no longer through rock blocks) virtually everywhere along its entire length.

In the final phase of the simulation, a little increase of the water level is sufficient in both models to obtain again excessive displacements of the head zone (no illustration). The models therefore suggest that soon after the critical event of 2007, this zone would be affected again by large movements, and that already in the short term, all the material above the main shear-zone of the head area ("head" in Fig. 36, page 81, i.e. the shallower shear-zone) would slide into the lower head zone or farther (compare Fig. 49). Thereby the foot of the crown zone is further unloaded and the crown zone becomes more unstable. So the movements along the deeper shear-zones of the crown and head area are increasing, bringing about considerable decrease in strength also there. In the years after 2008, the movements in the study area are therefore expected to deepen and to advance towards the slope crest, eventually also somewhat beyond.

At the time of finishing of this case study in summer 2009, important parts of these assumed consequences had already become reality: Critical accelerated movements of metres to tens of metres had affected the head zone at very short intervals: already in March 2008, in July after summerly rainfalls, in October 2008 and in January 2009 after a warm period with snowmelt. Again, these movements led to reactivations of the earth-flows further downslope. Although the upper head area had still not been eroded to the depth of the shallower shear-zone of the head area ("head" in Fig. 36, page 81), as it was expected (see Figure 49), the ground surface was not more than five metres above this level at that time. In the crown area, movements of several metres have produced a morphologic step of a height of more than three metres, destroying all four extensometers (compare Fig. 31, page 71). In the lower head area, an accumulation of material had formed, which is moving towards the steep step at the foot of the studied part of the slope and this accumulation is therefore expected to become unstable in the near future.

The large displacements that have occurred since 2001 have also brought about a lateral enlargement of the head area. In summer 2009, the area of the head zone was twice the size of that after the event of 2005. This effect is, needles to say, not reproduced by the 2D-models. However, the retrogressive evolution along the other margins of the head area is comparable to that along the investigated profile.

3.9. Updated geological, kinematical and mechanical understanding of the slope evolution

The different monitored displacement rates observed in different parts of the studied slope section during dormant, suspended and preparatory phases could be reproduced qualitatively by the models. Also the initiation of the critical phases and the rough spatial distribution of the displacements at their beginning could be explained logically by the simulations. The results suggest that the principal engineering-geological hypotheses taken as a basis for the numerical models are plausible. All in all, the reconstruction of the slope history in the simulations helped to test, to deepen and to improve the geological, kinematical and mechanical understanding of the evolution of the studied earth flow source area. According to this understanding, the slope evolution is the result of a complex interaction between changing material properties, changing hydrological conditions and deformations (see the scheme of Fig. 50).

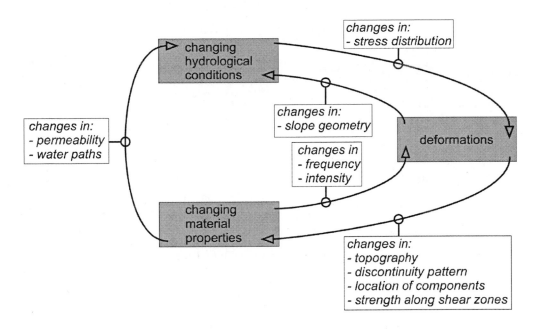

Fig. 50: Scheme of the complex interaction of different factors governing the evolution of the studied slope area.

Geologic and tectonic processes have predetermined the **material properties** and their alterability, as well as their susceptibility to weathering (Section 3.2., page 59). Together with the long-term climatic conditions they are determining the rate of weathering and the weathering grade (see the beginning of Section 3.5., page 65 and 66).

Short-term climatic variations, together with the geological composition and the morphology of the slope are governing the **hydrological conditions** and their variability (see Section 3.4., pages 63 to 65 and Section 3.7.1., pages 77 to 80). The changing hydrological conditions are linked to changes in the **stress distribution**. During the dormant, suspended and preparatory phases, a rise (or fall) of the water level, and thus of the pore pressures, causes a decrease (or increase) of the effective stresses.

In this manner, the changing hydrological conditions are controlling the **deformations** and their velocity. In the critical phases, in addition to this effect, significant changes of the *slope geometry* are taking place that are changing the hydrological situation, even if the climatic conditions and all the other external factors remain unaltered. Temporarily these changes in slope geometry are happening so fast that at some places in the slope they generate excess pore pressures, as a result of undrained-loading effects. These drastic changes of the water conditions bring about even larger deformations. Like all the other processes during the critical phases themselves, this feedback-effect was not part of the simulations, whereas the situation before and after each critical phase and the related intermediate geometry and load changes are modelled (see Section 3.7.3.3., pages 88 and 89).

With the above-described interrelation, the seasonal trends of accelerations and decelerations of the movements can be explained: the increase of activity in times of high pore water pressure, usually after the autumn months and in the spring months, as well as the timing of the critical events.

Yet in the monitoring period, repeatedly, climatic and hydrological conditions as they are typical for the beginning of the critical phases were observed without a reactivation of the landslide being triggered. For example, Figure 34 (page 75) shows that in the suspended phases 2006 and 2007, at not less than five points in time, the pore pressure measured in borehole B8B was higher than at the beginning of the critical event of 2007; in borehole B2B, this was the case at least three times. These observations demonstrate that the reactivations and accelerations of landslides of this type are not only controlled by climatically-driven changes of the pore pressure distribution.

It is furthermore of great importance that the material properties are changing with time and that these changes bring about changes in *intensity* and *frequency* of the deformations. This means that if the material properties deteriorate, acceleration phases become more frequent, they occur already with less important changes of the water level, are stronger and last longer. Definitely, these acceleration phases will continue following seasonal trends. The changing (deteriorating) of the material properties is a fundamental consequence of the geological composition of the rock masses of the study area, which are characterised by fast and deep weathering (Section 3.2., pages 58 to 60). And this fast and deep weathering is the precondition for the development of a landslide of this type at this place.

Inside the landslide, the changing of the material properties is strongly influenced –i.e. generally pushed on– by the deformations. In this decade, the changes in *topography* that are connected with the deformations have unloaded and thereby loosened important regions of the head and crown area. During the deformations, the *discontinuity pattern* of the materials is changing, for new fissures, cracks (e.g. tension cracks) and shear-zones are forming. The deformations also change the *positions of the very diverse components* inside the inhomogeneous materials, from blocky rock mass fragments to platy clay particles. This relocation and alignment causes a change – altogether this always means a decrease – of the shear *strength along the shear-zones*.

All in all, by local steepening of the topography, by loosening the rock mass with additional discontinuities, by the faster advancing of weathering along these discontinuities and by the decline of strength along the shear-zones, the head and crown area were destabilised more and more from event to event. This gives an explanation to the fact that the landslide activity has augmented in the investigated time span (see Fig. 29, page 64). It has to be assumed that this process of destabilisation has continued also after that time span and continues also at present (summer 2009), because the intervals between the critical events have decreased. Critical phases with movements in the metre-range and larger were observed in 2001, 2005 and 2007, three in 2008, and already one in 2009.

The changing of the material properties, especially of the discontinuity pattern, is linked to changes of the *permeability* and of the *water paths* inside the landslide body, and therefore influences the hydrological conditions. The effects of these changes are locally differing (compare Section 3.7.1., pages 78 to 79). In the crown area, the development of new cracks and fissures favours the infiltration and inside these discontinuities, where the permeability is much higher than in the surrounding rock, the water level can rise very fast and very far. In the head area, where the fluctuation of the water table is smaller and the permeability is lower, loosening and formation of additional discontinuities during the critical events cause a significant general drop of the pore water pressure (compare with the data of 2007/2008 in Section 3.6., pages 74 to 77).

The resulting stabilisation of the head area is the explanation for the slow-down and the calm phases directly after the critical events. Certainly these calm phases last only until the proceeding destabilisation of the crown area has again increased the ratio between driving and holding forces. This is why the critical event of 2001 was followed by a relatively long dormant phase. At that time, the destabilisation of the crown area had just begun; moreover, the crown was unloaded to a lesser extent than in the course of the next critical events. For this reason, the stabilisation of the head area showed its effects for a relatively long time. (Eventually, the landslide material in the head zone had even enough time to consolidate again slightly, so that locally, against the general trend, a re-increase of its strength might have occurred.)

3.10. Conclusions drawn with respect to the Valoria case study

Summing up, the reconstruction of the slope evolution by means of numerical modelling has led to an improved geological, kinematical and mechanical understanding of the crown and head area. The so obtained additional information is very valuable for the planning of measures intended to mitigate the risk of the Valoria landslide. Referring to the above-presented results of this work (compare also with the scheme in Fig. 50, page 113) the following statements can be made:

The deterioration of the material properties is a consequence of the geological composition and the tectonic history of the rocks themselves and therefore it cannot be stopped, as long as the rocks are exposed to any type of climate that includes humid seasons. In addition, because of the type of the landslide – i.e. landslide of the earthflow-type, presently appearing as a complex rock slide earth slide earth flow– and in the context of its history, further deformations, which again have a negative impact on the material properties, must always be expected.

Therefore, if eventually planned active mitigation measures are to be successful in the long term, they must effect a significant and durable lowering of the pore water pressures in the crown area, especially in the still inactive adjacent areas of possible enlargement of the head zone. And these mitigation measures have to withstand a maximum of deformations. In any case, it will be very difficult to meet these demands. In contrast, a short-term success (within a few years) of any kind of measures is unlikely, because it is very probable that not all the consequences of the destabilisation caused since 2001 by the repeated critical events, have come to light by now.

Independent of considering eventual active mitigation measures, passive mitigation measures, as they were undertaken and extended since 2001 (especially continuous monitoring), are indispensable. This is particularly because the neighbouring areas (with respect to the actual landslide areas), which are partly inhabited, mainly consist of old earthflow deposits and could relatively easily be mobilised in the course of future critical phases. Also in the future, mitigation measures must include a continuous monitoring of the movements and of the hydrological conditions, in order to recognise new critical events in time, and to warn the residents that are concerned. This is of particular importance because further reactivations of the entire complex earth slide/flow must be expected and are possible at any time. These reactivations can even be of a larger extent than the previous ones, for large parts of the crown area, which were unloaded and oversteepened in the last years, could collapse. In any case, the retrogression of the crown zone will proceed further. And in addition, a large amount of earth-slide material has accumulated in the lower head area, moving towards the steep step in the transition zone between the upper source area and the lower source area.

4. Geotechnical and numerical modelling for the pre-evaluation of mitigation measures in the source area of the Corvara earthflow applying an Inverse Parameter Identification technique

4.1. Short introduction to the Corvara case study

Located in a renowned tourist area in the Dolomites of Italy, the Corvara landslide (see Fig. 51) was selected for its socio-economic relevance and for the availability of an extensive dataset of field investigation and monitoring data (*Corsini et al. 2005, Panizza et al. 2006, Corsini 2009 U*).

Fig. 51: Location and overview of the Corvara earthflow [complex earth slide earth flow] (slightly modified after *Corsini 2009 U*).

It extends from the Col Alto to Pralongià, stretching in altitude from about 1600 m a.s.l. to around 2100 m a.s.l. With an estimated 30 million cubic metres of overall displaced debris, this active slow-moving rotational and translational earth-slide/earth-flow damages a national road and a set of facilities including ski infrastructures, electricity lines and a golf-court. Moreover, in worst case scenarios, the landslide might affect some buildings located in front of its toe and possibly endanger the downstream settlements by damming up the streams running at its flanks (*Panizza et al. 2006*).

For this reason, a complete geological, geomorphologic and geotechnical analysis of the landslide was carried out since 1996, which was intensified in the period between 2000 and 2005, thanks to the support of the Autonomous Province of Bolzano - South Tyrol, the main local stakeholder, together with the Corvara municipality (*Corsini et al. 2005, Panizza et al. 2006, Corsini 2009 U*). The study involved drilling and retrieving more than 500 m of cores, installing ten inclinometer casings, also TDR-cables (Time Domain Reflectometry for detecting shear zone deformations, see e.g. *Dowding et al. 2001*), and half a dozen electric piezometers, furthermore, the setting up of a complete on-site weather station, the performing of around 5 km of geophysical profiles (refraction seismics combined with resistivity) and finally, deploying all over the concerned slope areas about fifty GPS (Global Positioning System)- benchmarks. The main outcomes were a precise geological and geomorphological mapping and geotechnical characterisation of the landslide, a reliable estimate of the volume of the clay-rich debris involved in movements, a reconstruction of the Holocene evolution of the landslide and a measure of the movement rate and depth in different landslide sectors.

Regarding this latter point, it is worth anticipating that the average rate of movement over sliding surfaces that are 20-40 metres deep, ranged in the monitored period from cm to m per year, with the highest rates in the uppermost part of the accumulation lobe and in the track zone (*Corsini et al. 2005*). The above-described investigations led to the identification of landslide sectors that are particularly active or of interest with respect to future risk scenarios. This understanding of the phenomenon was used to lay out risk zones and to propose a set of options for undertaking focussed slope stabilisation works in these critical sectors of the landslide (*Panizza et al. 2006*). Among these sectors, source area S3 was found to be considered primarily for eventually planned mitigation interventions and therefore was chosen for testing the numerical modelling concept presented in this work.

4.2. Basic field data

4.2.1. Geology of the landslide area

The geological features of the region have been studied in detail by various authors since the beginning of the 20th century (for literature references see *Corsini et al. 2005*). The landslide slope is made up of Triassic flysch-type rock masses characterised by a semi-regular alternation of sub-metre thick strata of arenites or marly limestones, which represent the harder components, and sub-metre to metre thick packets of finely laminated claystones or clayshales, very weak rocks that are partly of volcano-sedimentary origin. Bedrock forms a monocline dipping upslope at about 30° of inclination, is in right-up polarity and shows clear evidence of tectonics-related joint sets linked to Alpine orogenetic phases (*Corsini 2000*). The transition from the older La Valle Formation, in which the harder rock component is made up of arenites, to the younger San Cassiano Formation, in which the harder components are made up of marly limestones, occurs immediately down-slope from the landslide crown. Therefore, the bedrock of the landslide area is actually mostly composed of La Valle strata, in which volcanoclastic arenites can be relatively hard, and their ratio to the weak rock component, mainly claystones, varies from 1 to more than 2 (*Corsini et al. 2005*).

Fast and deep weathering represents an essential feature of the bedrock formations; this is why they have developed extremely thick soil covers. The thickness of these clay-rich layers reaches from metres to tens of metres in autochthonous positions and up to 100 metres in semi-autochthonous and allochthonous positions. Over the whole Quaternary period, with interruptions in cold stages, they have been frequently involved in movements and have been the cause of many landslides.

4.2.2. Local Geomorphology and quaternary geological history of the slope

The overall geomorphologic setting of the slope is primarily controlled by geologic structures and secondarily, by the shaping action of Pleistocene glaciers and of Holocene mass wasting processes (*Corsini et al. 1999, Corsini 2000*). On the basis of till deposits distribution, it was estimated that the whole slope was covered by an ice cap during the last Würm maximum. Its absolute elevation was probably in the order of 2200 - 2300 m a.s.l. Therefore, the thickness of the ice cap on the landslide slope ranged from about 100 metres in the actual upper source areas to 900 metres in the present-day foot zone of the accumulation lobe. The glacial history has induced load stress into the slope and has caused a quite pervasive permafrost layer to develop in the upper part of the bedrock (*Corsini 2000, Corsini 2009 U*).

Actually, several radiocarbon datings indicate that the landslide has started affecting the slope from about 10000 cal BP. It is reasonable to speculate over a direct link between the failure of the slope, its de-buttressing after deglaciation and permafrost melting at the end of the Lateglacial (*Soldati et al. 2004*). A second recognised major phase of activity of the landslide occurred between 5000 and 2500 cal. BP, possibly in conjunction with a worsening of the climatic conditions during the Holocene (*Corsini et al. 2001, Borgatti et al. 2007 A, Borgatti and Soldati 2009*).

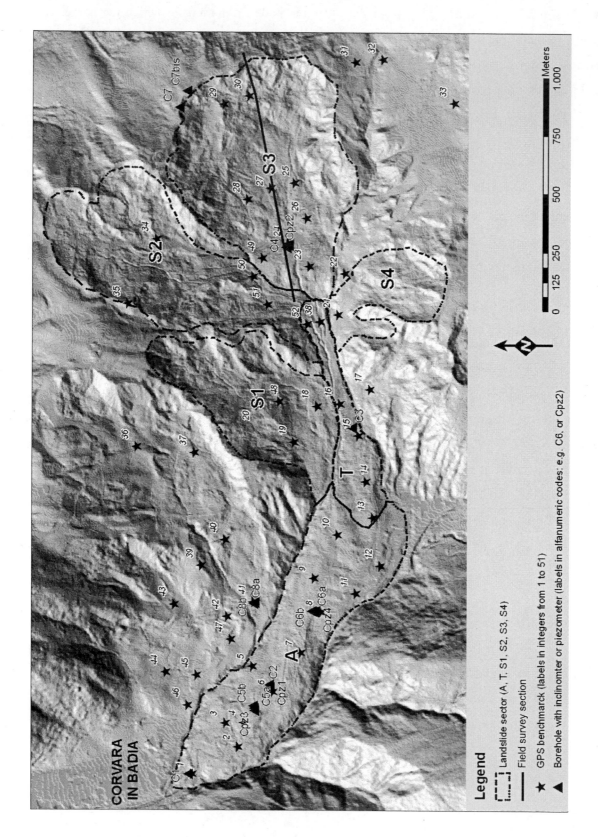

Fig. 52: Present-day slope morphology, boundaries of the landslide sectors denominated in Figure 51 (page 117), measurement points and location of the field survey section of Figure 56 (page 128) (slightly modified after *Corsini 2009 U*).

At present, the Corvara landslide has developed to the point at which well distinct source (S), track (T) and accumulation (A) areas can be outlined (see Fig. 52). The source area itself can be subdivided into four sectors (S1, S2, S3, and S4). Present-day movements are "very slow" to "slow", according to the classes proposed by *Cruden and Varnes 1996* (see Section 4.2.4., page 122, for details on the displacement rates). Nevertheless, observations made in borehole cores and radiocarbon data point to a probably differing landslide development during the Holocene that was characterised also by more pronounced periods of increased activity, during which series of reactivation events caused earth-flows to accelerate and stack over older landslide debris. During these periods, the distribution of activity was most likely retrogressive at the crown and at the flanks of the source areas, enlarging at the sides of the accumulation area, advancing at the turn of the landslide into the main valley and at the toe (*Corsini et al. 2005, Panizza et al. 2006*).

4.2.3. Specific relevance of Source Area S3 for future risk scenarios

In the last decade (and referring especially to the time span of the intensified monitoring campaign 2001-2004), signs of retrogressive activity were observed at the upper end of the track zone, in which the highest displacement rates were measured. This causes erosion in the foot zone of source area S2, which is at present (2009) the most active source area, and of source area S3 which is at the moment not discharging landslide debris into the track zone (T) of the landslide. In practice, thanks to the presence of a buried cuesta-type rock ridge extending transversally to the overall slope at the lower end of source area S3, all the landslide debris accumulated in this source area, which represents the largest source area, is tied back.

Considerable displacement rates in source area S3 taking place throughout the whole year and with seasonal acceleration phases (see next Section) clearly show that this situation is not the result of a strong support in the foot zone but rather of a weak equilibrium defined by the gravitational forces of the landslide masses, their frictional resistance and the underlying bedrock topography.

Therefore, given the scenario of climate extremes and/or an excess pore pressure zone created by sudden major roto-translational soil slips or slumps in the steeper upper parts of the slope, or by a local speedup of shallower earth-flow lobes in its mid part, accelerations of the movements of source area S3 must be expected. This would imply a large extra aliquot of landslide debris to reach the track zone, with movements propagating to the point at which the whole landslide accumulation farther downslope towards the village of Corvara could also be displaced at a velocity higher than the presently monitored one. Damming of the side-flank streams, Rio Rutorto and Rio Chiesa, and disruption of the national road running over the landslide accumulation area would be the most likely consequences of this chain-effect. Therefore, countermeasure works that could prevent a possible worsening of stability conditions in this sector are of particular interest.

121

4.2.4. Contemporary displacement rates of Source Area S3

Displacement rates in landslide sector S3 are properly represented by D-GPS (Differential Global Positioning System) data from benchmarks 23-30 (see Fig 52, page 120). These data refer to RTK (Real Time Kinematic) processing of monitoring campaigns carried out four to five times a year in the period from 2001 to 2004 (for details on the monitoring techniques see *Corsini et al 2005* and *Caroli 2009*). The displacement history of these benchmarks is plotted in Figure 54 (page 123), showing the norms of their 3D displacement vectors for each measurement date whereas Figure 53 displays the directions of these vectors in plan view.

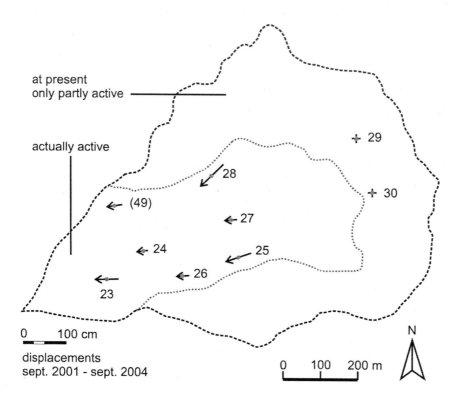

Fig. 53: Displacement vectors of GPS-benchmarks 23-30 between 2001 and 2004 in plan view (GPS data from *Panizza et al. 2006* and *Corsini 2009 U*). Benchmark 49 was measured starting from 2003; vector has been scaled up respectively.

Over the triennial period of intensified measurements, these values exhibit approximately linear trends, which (except for benchmark 28) were also confirmed by the ongoing less frequent measurements until 2008 (*Caroli 2009*).

GPS points 29 and 30, located in a relatively stable area below the crest of the slope and above the actually active part of source area S3, did not show significant displacements. After three years, the obtained values were still below (benchmark 29) or near (benchmark 30) the range of the measurement error of the utilised GPS-method, which ranges around one centimetre for horizontal and circa five centimetres for vertical displacements (*Corsini, A., University of Modena and Reggio Emilia, personal communication*).

122

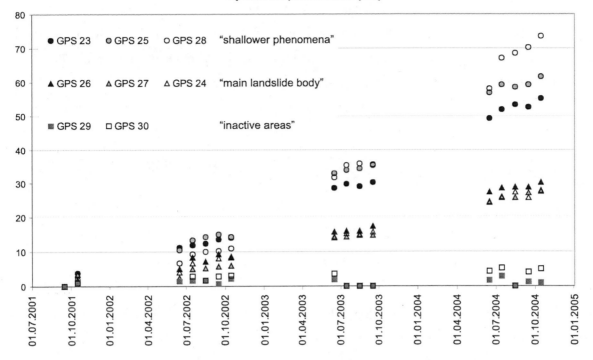

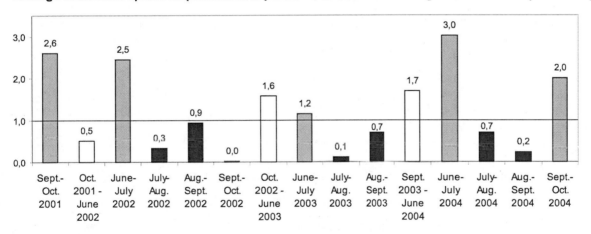

Fig. 54: Displacements measured in source area S3 between 2001 and 2004

(GPS-data from *Panizza et al. 2006* and *Corsini 2009 U*).

Rates of around one decimetre per year were measured at benchmarks 24, 26 and 27, lying on the main earth-slide body of source area S3. Significantly higher displacement rates, of around two decimetres per year, were observed in areas where, in addition to this movement of the main earth-slide body, the slope is also affected by shallower earth-slide and earth-flow phenomena (GPS points 23, 25 and 28). The displacements of benchmarks 23-28 reveal seasonal variations, which appear to be more pronounced in the case of the last-mentioned group of benchmarks. These variations are to be perceived as delayed reactions of the landslide, at any one time, to the climatic conditions of the recent weeks to months rather than to single meteorological events (*Corsini et al. 2005*).

123

Measurements were always carried out on (or around) the fifteenth of each month and, for reason of accessibility to the benchmarks, between June and October, alternatively between June and September. Therefore they primarily capture the season of the alpine summer, which on the landslide site lasts usually from July to September, depending on the climate of the year. This time is characterised by low displacement rates, slower than the average trend (brown columns in the lower part of Figure 54), because then the strong insolation during the (astronomical) summer, which is linked to higher temperatures, increased evapo-transpiration as well as the concentration of rainfalls in shorter time intervals, and therefore to a reduced infiltration despite high precipitation, shows its (delayed) effect.

Towards autumn, evapo-transpiration decreases with the duration of the precipitation events increasing; while still considerable amounts of rain and ephemeral snow are falling (climate data for the modelled period are given in Fig. 62 on page 140). As a result, in the time following this season, as well as in the alpine spring (usually May and June on the Corvara landslide), when, deeper inside the slope, the water of the past snowmelt – normally taking place in April – adds to that of the precipitations, elevated displacement rates are recorded (columns in light blue in Fig. 54).

The displacement rates calculated for the intermediate intervals between the monthly measurements (colourless columns in Fig. 54) are reflecting – in varying parts – both the increased values of spring and autumn as well as the low rates of the winterly calm phase, resulting from the temporary stop of infiltration due to frozen ground and snow cover.

4.3. Geometry Model of Source Area S3

Evidence from boreholes drilled along the main longitudinal axis of the entire landslide, coupled with an interpretation of geoelectric and seismic refraction data, has actually shown that the thickness of the landslide materials is largest in the upper part of the accumulation area (about 90 m) and is elsewhere mostly ranging from 10 to about 50 m *(Panizza et al. 2006)*. The clayey landslide debris is rather loose in the upper 10 to 20 m and more compact at greater depths. In some borehole cores, metre-thick levels, in which landslide debris shows a reduced matrix fraction, mark the interface between stacked older and younger landslide deposits. Inclinometer data have shown that, at present, these younger-older landslide deposits interfaces are among the main sliding surfaces *(Corsini 2009 U)*.

Input data for setting up the geometry model of source area S3 include surface topography, subsurface boundaries (i.e. the subsurface geometry), depth of sliding and dimensions of the different landslide bodies characterised by different displacement rates. Surface topography has been obtained by profiling a DEM (Digital Elevation Model) produced in 2005 by the Autonomous province of Bolzano using airborne LIDAR (Light Detection and Ranging). Its nominal elevation accuracy is in the order of one metre. Subsurface boundaries, including the depth of sliding surfaces, have been obtained by interpreting field evidence, borehole stratigraphy, seismic refraction data, and geoelectric and inclinometric measurements.

4.3.1. Ground surface of the 2D Geometry Model

The DEM resulting from the LIDAR measurements allowed for a very exact determination of the 3D-Geometry of the ground surface of landslide source area S3. Based on the so obtained detailed morphology and precise slope inclination data, the slope of source area S3 was divided into several morphological units based on engineering-geological considerations (see Fig. 55).

Eight sections were drawn across source area S3 in the direction of the detected movements and longitudinally to the overall slope. In doing so, the slopes that are related to the intermediate ridges between the different neighbouring source areas and slopes that are oriented perpendicular to the overall slope of source area S3 and to the measured movements, were excluded. These morphological features can not be reflected by a 2D Geometry model. Wherever ridge and hollow positions were found alternating perpendicular to the overall slope, they were traced alternately also by the different sections, in order to obtain a representative average slope inclination in these areas. The course of all eight sections was further defined in such a way that fluvial features were never followed longitudinally but always cut more or less perpendicularly or at least transversely. This was done because, naturally, fluvial processes are not part of the model, and the changes in morphology caused by the little streams on the landslide body, which are locally quite remarkable, would elsewise bring about a misleading levelling of the surface topography of the geometry model.

The sections were then averaged for each morphological unit. For this purpose, in a GIS-system, the elevation given by the DEM was sampled along each section at a sufficiently large number of equidistant points. The number of sampling points was set equal for all eight sections through a respective morphological unit and in such a way that for the longest section through this unit, the distance between the sampling points was still smaller than the raster resolution of the DEM (< 2.5 m).

The locations of field instruments were projected from their 3D positions on the ground surface into the 2D model by assigning them in each case to the nearest sampling point. Finally, by averaging distances and altitudes, an average surface section for the geometry model of source area S3 was obtained.

125

Division of source area S3 into different morphological units

Hillshade

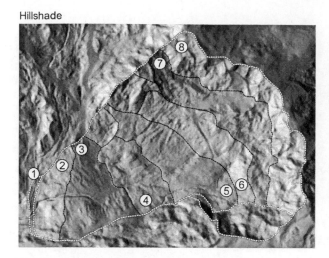

Slope inclination

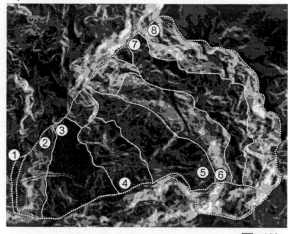

① Flat transition zone between S3 and downslope landslide sectors (S2 and T)

② Steep front of landslide block C4

③ Top of "Block C4", flat, with ponds

④ Gently inclined zone with shallower earth-flow lobes, downslope part

⑤ Gently inclined zone with shallower earth-flow lobes, upslope part

⑥ Steep zone with "unsteady" morphology

⑦ Plateau formed by old sliding-bodies

⑧ Headscarp, steepest zone of the slope

■ <10°
■ 10-20°
■ 20-30°
□ 30-40°
□ >40°

Averaging of 8 topography sections for each morphological unit

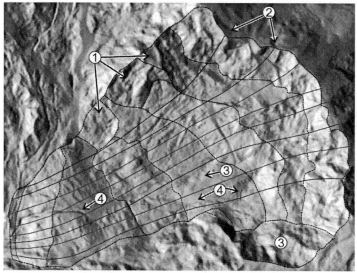

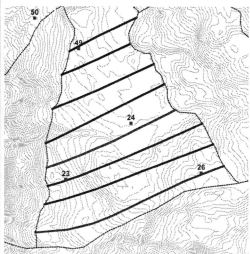

① Slopes related to intermediate ridge between source areas S3 and S2: excluded

② Ridge-positions and hollow-positions: traced alternately

③ Slopes perpendicular to overall slope and movements: excluded

④ Fluvial features: never traced longitudinally

- Equal number of points for each section part
- GPS-benchmarks assigned to nearest section points
- Maximum distance smaller than 2.5 m

→ Representative surface section by averaging distances and altitudes

Fig. 55: Deriving a representative surface section for the 2D model of source area S3 from LIDAR data (LIDAR raster data from *Corsini 2009 U*).

4.3.2. Sub-surface components of the Geometry Model

Core profiles of boreholes C4 and Cpz2, located at the lower end of landslide sector S3 (see Fig. 52, page 120), were particularly important for defining subsurface boundaries within the landslide material. Borehole C4, instrumented with an inclinometer, allowed defining: a first inner boundary at 43 m depth, corresponding to a change of debris compaction and to the main sliding surface and a second boundary at 54 m depth, corresponding to the contact between landslide debris and weathered bedrock. These boundaries were extended upslope and into a 2D section on the basis of a detailed survey in the field along a representative profile through the central part of source area S3 (see Fig. 56 for the profile and Figure 52 on page 120 for the location of the survey section), which was then combined with the analysis of seismic refraction data and a good deal of expert-knowledge interpretation of geomorphic break-lines in the slope. Thereby the author had recourse to the experiences obtained by the Applied Geology research group at the University of Modena and Reggio Emilia during the field monitoring of the Corvara landslide and several similar landslides of the earthflow-type (e.g. *Corsini et al. 2005, Borgatti et al. 2006, Ronchetti et al. 2007, Corsini et al. 2009,* or *Ronchetti et al. 2009 A*).

By documenting signs of movement, disintegration or segmentation in the field, important evidence for the location of detachment zones and coherent areas lying in between them, as well as more stable areas, could be gathered. Borehole C4 also allowed detecting a third inner boundary at about 30 metres, corresponding to a change of debris compaction and to a presumed onset of internal creep deformations of the landslide body, which are superimposed to those of the main shear-zone. As geophysical data and geomorphic field evidence do not allow for interpreting this boundary upslope unambiguously and the internal deformations presumably associated with this zone are about (at least) one order of magnitude smaller than those occurring in the main shear-zone, this boundary was not included into the geometry model.

The schematic cross section derived from all this information is shown in Figure 57 (page 129). Currently, the main detachment areas of the earth-slide mass of source area S3 are located at around 2000 m a.s.l at the lower edge of a slightly inclined plateau, which, situated at the foot of the morphological headscarp of the landslide, is built up of old sliding bodies. The area below this main detachment zone until the slope-foot of source area S3 is moving as one coherent layer along a basal shear-zone 20-40 metres deep. Further detachment zones within this part of the slope can be recognised in the field (see also Fig. 56) due to the morphology of the ground surface, but they appear not to be active at present and already for a longer time (some years or decades). When discussing future hazard scenarios, these detachment zones are definitely a matter of particular interest, because they could easily be reactivated. But this does not hold true for the modelling of the situation in the years of the monitoring campaign. As a result, they are not part of the geometry model.

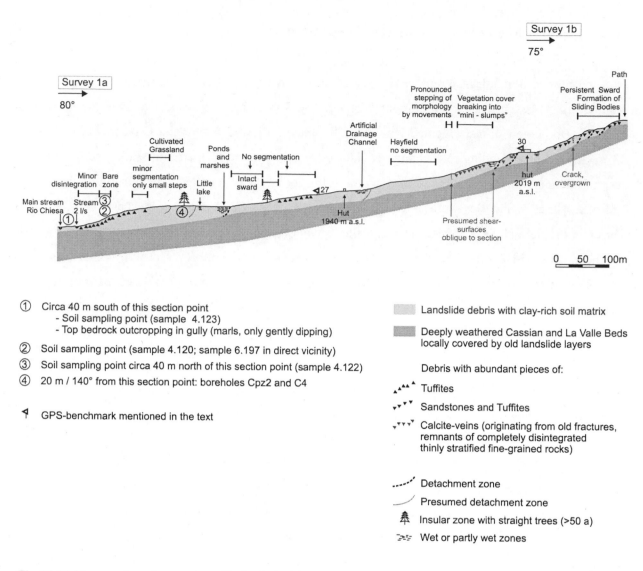

Fig. 56: Field survey through source area S3; for the location of the section, see Fig. 52 (page 120).

Upslope from the main detachment-zone, the basal shear-surface near the top of the bedrock is at present not fully developed. Signs of slope-movement, though present, are by now becoming overgrown by the vegetation. Below the crest, the thick soil cover has dismembered into separate sliding bodies. At the moment, the detachment of these blocks is taking place only locally, predominantly in the form of sudden soil slips or slumps affecting only relatively narrow strips of the slope, rather than in the form of continuous creep displacements of the earth-slide or earth-flow type.

In the foot area of source area S3, in the transition zone towards source area S2 and towards the track zone T (see Fig. 52, for the location of these areas), the examination of morphological indicators (cracks, steps, bulges etc.) of compression and dilation suggested that the latter are rather prevailing. This means that the neighbouring downslope soil bodies are currently moving or just along or even away from source area S3 and do not exert a considerable support onto it. Furthermore, the geophysical measurements and an outcrop, visible in a gully in 2004 (compare Fig. 56, point 1), suggest that the soil cover above the stable bedrock thins out to a thickness of only a few metres in this transition zone. The geometry model is interpreting these observations in the way that – due to the

128

buried rock ridge mentioned in Section 4.2.3. (page 121) – the basal shear-zone of source area S3 is slightly dipping upslope in the foot area and cropping out before reaching the transition zone. Under the basal shear-zone, the slope is considered to be stable. In the lower part, this zone is made up of older landslide material which has been trapped in a hollow. In the upper parts, it is composed of bedrock layers of the formations described in Section 4.2.1. (page 118).

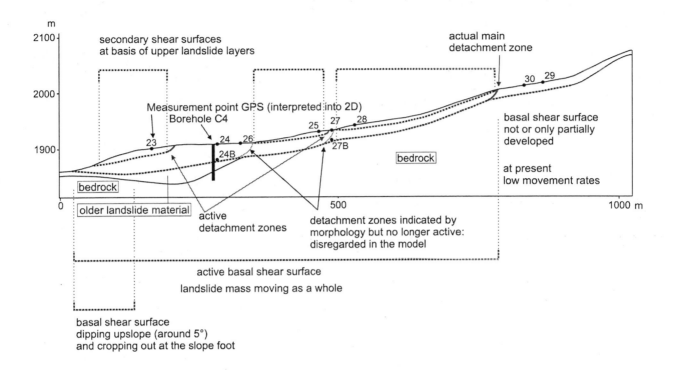

Fig. 57: 2D Geometry Model of source area S3.

Near the ground surface – expressed by a bulgy and hummocky surface morphology and visible earth-flow-type lateral ridges in the field – the slope is regionally affected by shallower earth-slide and earth-flow phenomena. These are reaching between 3 and 10 metres deep, and are concentrated in two areas, an upper and a lower area, between the actual main detachment region and the flat, partially inversely inclined area around borehole C4. At the basis of these upper landslide layers, the geometry model is assuming secondary shear-zones, same as in the oversteepened front area of the so-called block of C4, where the landslide material has been piled up.

4.3.3. Internal stratification of the Geometry Model

Figure 58 describes the internal stratification of the slope in the Geometry Model and explains how it was derived based on the observations made in borehole C4. Similar core profiles and deformation patterns have been encountered in all the other boreholes across the Corvara landslide, which exhibited deformations (see Fig. 52, page 120, for the location of the boreholes): in boreholes C2, C5 and C6, probably also C3, where the tube broke after less than one month, being measurable only once more above this rupture point. Also the displacement profile of borehole C8, belonging to a neighbouring landslide of a similar type, bears resemblance to that of C4 (*Corsini et al. 2005, Panizza et al. 2006*). Therefore this stratification scheme was applied to the whole geometry model of source area S3.

The displacement measurements in borehole C4 could be carried out only for one month along the entire length of the inclinometer tube, and then the tube was destroyed in the range of the main shear-zone. However, this time-span was sufficient to identify a clearly discernible displacement-profile, proving that the deformations were concentrating on this relatively thin shear-zone. The region above this zone shows little internal deformations. These could be measured regularly until the depth of 40 metres between summer 1999 and spring 2004, averaging less than one centimetre per year. The total displacements of the ground surface at this point, measured by GPS between 2001 and 2004, added up to nearly 10 cm per year in this time interval. In Figure 58, both inclinometer measurements were brought to the same scale assuming that the sum of both profiles results in the displacement rate obtained by GPS at the surface.

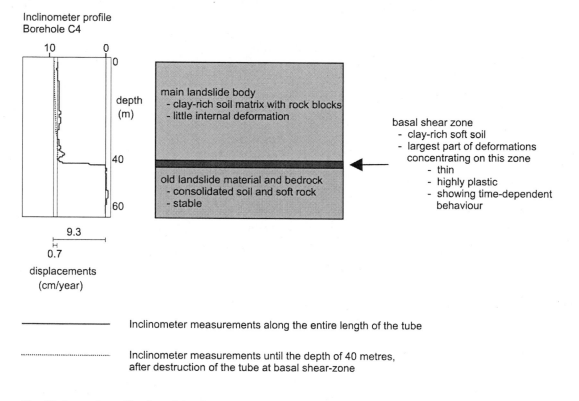

Fig. 58: Internal stratification of the Geometry Model (Inclinometer data from *Corsini et al. 2005* and *Panizza et al. 2006*).

From the inclinometer measurements and deformation measurements by means of TDR cables it can be inferred that the thickness of the shear zones is definitely less than 2 metres. *Corsini et al. 2005* estimated that this value is generally in the order of 1-2 m. As the exact thickness of the shear-zones is

130

not known, and certainly varies for the different shear-zones and also for different shear-zone domains, a fixed thickness of 2.0 m is used for all shear-zones of the geometry model.

4.4. Material Models

The landslide body is made up of a soil matrix of silty clays and clayey silts, which encloses coarser components, such as a variety of gravel-size particles and rock blocks of a size of up to several metres. These consist of volcano-sedimentary arenites and calcareous sandstones, marly limestones or dolostones. The coarser components, which are mainly of angular shapes, i.e. they show little rounding for they have not undergone considerable sedimentary transport, are mostly floating in the clayey soil matrix and do not support each other. This may not hold true really everywhere in the landslide body without limitations, but for the mechanical behaviour of the shear-zones, which are assumed to pass through its weakest parts, only the properties of the clay-rich soil matrix are considered to be relevant.

Depending on whether the landslide is overriding the bedrock or older landslide deposits, the regions under the basal shear-zone are composed of heavily weathered clayey and silty soft-rocks or a relatively compact layer of clay-rich soil with coarser components and rock blocks. These older landslide deposits have formed analogue to the actual landslide material but have consolidated with time. By means of Radiocarbon analyses, *Corsini et al. 1999* dated the beginning of the landsliding activity of source area S3 to 2500-3000 years before present. When assigning material properties to the different parts of the geometry model, both these materials were treated as one, because the altered bedrock layers and the compacted landslide materials gave very similar responses during the geophysical soundings (*Panizza et al. 2006*) and therefore separating them would not increase the precision of the model but just introduce additional uncertainties.

Based on the above considerations, the soil along the shear-zones can be regarded as a highly plastic material which is characterised by a pronounced time-dependent behaviour. Therefore, the Soft Soil Creep Model (*Vermeer and Neher 1999*) was chosen as a constitutive model for these zones. For the description of this model and its model parameters, please see again pages 85 and 86 in Chapter 3. Due to the age of the earth-slide/flow activity in source area S3 (see above), and the continuous nature of the sliding processes even in the present dormant phase (see Section 4.2.4., page 122), modelling of the shear-zone layers was based on the following two hypotheses.

First: Although the positions of the shear-zones are known to change from time to time in different domains as a reaction to the changing slope morphology, it is assumed that the actual shear-zones, in their largest parts, have undergone relative displacements at least in the range of tens of metres. Therefore the material of the shear-zones is hypothesised to have reached a characteristic minimum (residual) strength which does not decrease significantly any more with additional deformations. Second: As long as neither stress-distribution nor slope geometry are changing, displacement rates along the shear-zones are assumed to be constant. The variations of displacement rates described in Section 4.2.4. (page 123) are therefore linked exclusively to changing pore water pressures.

With respect to material modelling, the detachment zones were treated as parts of the respective shear-zones and so the Soft Soil Creep Model was used also in these sub-zones of the geometry model. The blocks and layers below and above the shear-zones show little internal deformations, which are relatively small compared to those of the shear-zones. For this reason, their material behaviour is not modelled as elasto-plastic. However, it is necessary to account for the gravitational loads that these blocks and layers exert onto the shear-zone layers and a rough estimate of their stiffness is used for simulating to which extent they are capable to transmit stresses and deformations. Therefore a linear-elastic material model is used for these blocks and layers.

4.5. Hydrogeological Model

As part of the numerical modelling concept of this work, a hydrogeological model describes the distribution of pore water pressures inside the slope with time. The set-up of this hydrogeological model comprised numerical simulations of groundwater flow. As the inverse modelling technique of *Meier et al. 2008,* which is used in this work, was not operational for its application to groundwater models at the time when this study was carried out, the calibration of these groundwater flow models was done by means of the traditional trial-and-error method.

On principle, inverse modelling methods lend themselves and are recommended and planned to be used also for the calibration of the hydrogeological model. Applications of inverse modelling in groundwater flow simulations have been presented since many years by numerous authors (e.g. *Poeter and Hill 1997, Samper-Calvete and Garcia-Vera 1998, Finsterle 1998, Finsterle 2000, Carrera et al. 2005*). However, such applications were not treated in this work, whose focus is on deformation modelling with the feature of the hydrogeological modelling being used as fixed input.

4.5.1. Basic hydrogeological data

As a consequence of the dimensions of the Corvara landslide and the heterogeneity of the slope with its diversified morphology, the piezometer measurements (*Panizza et al. 2006*) are able to provide only punctual information on the hydrogeological conditions. Only three of the eight piezometers installed at the landslide site (the instruments of boreholes Cpz2, C4 and C7) lie within source area S3 (see Fig. 52, page 120, for the location of the boreholes). Measurement series recorded by these piezometers revealed the existence of two different types of overlapping groundwater regimes whose influence varies locally inside the slope: one that is connected to the hydrographic pattern, e.g. nearby streams, ponds and marshes, and therefore shows relatively small variations (Type A) and one that is linked to the infiltration of precipitation water and consequently is undergoing much larger seasonal fluctuations (Type B). Due to the low permeability of the clay-rich landslide materials themselves, these water pressure regimes are acting chiefly along the discontinuities of the landslide mass which consist in numerous tension cracks, shear cracks and – only near the ground surface – also desiccation cracks (compare also Chapter 2, page 38 to 40).

Therefore, at a local scale, changes in the pattern of discontinuities that are produced by the movement of the landslide can lead to considerable changes in the interaction of the different groundwater regimes. This was already presumed at an early stage of the monitoring campaign, when in June 2000 the piezometric measurements recorded a sudden uprise of the average water table in borehole C4 (by around 6 metres), which involved a reduction of the seasonal fluctuation range (from more than 4 to less than 2 m) and appeared to be irreversible in the following years (*Corsini et al. 2005*). While the fluctuations can be attributed quite well to precipitation events, an enhanced hydraulic connection to the constant high pore pressure regime of the nearby ponds and marshes is regarded in this work as the cause of the long-lasting rise of the average water level as well as for the decrease in fluctuation range. In order to fix hydrogeological boundary conditions not only for the locations of the boreholes but for the entire 2D slope model of source area S3, a series of assumptions had to be made (see Fig. 59). These are based on field mapping of springs, streams and ponds, and of areas of elevated humidity, such as marshes, which was carried out in 2002 (*Panizza et al. 2006*), and on the field survey of autumn 2004 (see Fig. 56, page 128). Given the uncertainties underlying these assumptions, and the resulting impracticality of a more detailed description, the hydrogeological conditions observed along the slope were idealised to a single continuous aquifer marked by one average hydraulic level.

4.5.2. Definition of a simplifying continuous average water level

The level of pore water pressure (see Fig. 59) was assumed to be found permanently near the ground surface at the slope-foot of source area S3. Its average depth in the front accumulation was taken from the measurements in boreholes Cpz2 and C4, where it varied between one and eight metres. In the adjacent flat ponding area, the water level was assumed to touch again and permanently the ground surface. Further upslope, in the area of the shallow secondary earth-slide/earth-flow lobes, it remains high, attaining a greater depth in the convex main detachment region. In the nearly flat neighbouring marsh area, the water level reaches again the ground surface, while its depth in the crest region was given by the results of the measurements in borehole C7, where the obtained values ranged between three and eight metres. If a constant average water level and a constant pore pressure distribution are used as input for the deformation model, then only constant displacement rates are calculated for all time spans in which the slope geometry does not change significantly (compare Section 4.4., page 131 and 132).

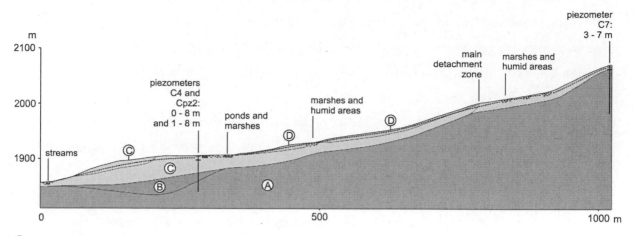

Fig. 59: Simplified hydrogeological model assuming one continuous average water level
(piezometer data from *Panizza et al. 2006*).

4.5.3. Definition of two different hydrogeological scenarios

Although the measured displacements can be approximated quite well by linear trends, their rates are not constant but vary seasonally and according to the climate of the year, e.g. displacements between October 2001 and October 2002 were smaller than those observed between October 2003 and October 2004. These climatically-driven variations can only be simulated realistically by assuming a time-variant pore water pressure distribution. Owing to the observations depicted in Section 4.2.4. (pages 122 to 124), the hydrogeological model assumes a low-pore- pressure scenario for periods of low infiltration, in summer and in winter, whereas for phases characterised by high infiltration, in times of frequent precipitation in spring and in autumn as well as after snowmelt, a high-pore-pressure scenario is taken as a basis for the calculation of this distribution. In addition, a change in infiltration conditions is followed by a transition period of defined duration, in which pore water pressures are gradually changing until at the end, the new scenario fully takes effect throughout the landslide.

4.5.4. Permeabilities of the materials involved

The deformation model will take into account the transient pore water pressures following the theory of consolidation. Therefore, the permeability parameters of the materials have an effect both on the (spatial and temporal) distribution of pore pressures during the rises and falls of the water pressure level and on the calculated deformations.

4.5.4.1. Basic assumptions concerning the permeability parameters

Average permeability values were assigned to the different layers of the geometry model, which were chosen within the value ranges reported by other authors from similar landslides that have developed in a comparable geological setting (e.g. *Bonomi and Cavallin 1999*, *Malet et al. 2005*, or *Simoni et al. 2004*). These are also in good agreement with the results of field tests carried out by the Applied Geology research group at the University of Modena and Reggio Emilia on complex earth slides earth flows in the Apennines (e.g. *Ronchetti et al. 2009 A* and *Ronchetti et al. 2009 B*).

In order to simplify the hydrogeological assumptions, isotropic permeabilities were set for all materials but an exception was made for the two upper landslide layers, assuming that due to their lower degree of compaction and elevated movement rates, they are characterised by a higher number of discontinuities. In the case of landslides of the earth-slide and earth-flow type, near the ground surface, vertical discontinuities prevail over the horizontal ones and therefore the vertical permeability of these layers was assumed to be significantly (here: three times) higher. The region above the secondary shear-zone in the front area is considered to be made up by materials that are less intensely weathered than the upper landslide layers (D in Figure 59, page 134) and to consist of the same material as the rest of the accumulation of Block C4 and the main landslide body (C in Figure 59, page 134). Therefore it was not modelled more in detail. With respect to the main landslide body, for which a value of 1 E-7 m/s was set – this represents a standard value corresponding to the literature references above cited – the older landslide material, due to long-term consolidation (see Section 4.4.), has a somewhat lower hydraulic conductivity, while the permeability of the bedrock, although heavily weathered, is assumed to be still significantly smaller, and was therefore fixed at a value one order of magnitude lower than that of the main body.

4.5.4.2. Steady-State Groundwater Flow Calculation as plausibility test, and adjustment of permeability parameters

In order to test the plausibility of the assumed permeability parameters and to adjust their values for a most realistic approximation of the real field conditions, a steady-state groundwater flow calculation based on the Geometry Model was carried out in cooperation with the Applied Geology research group at the University of Modena and Reggio Emilia (*Ronchetti 2007 U* and this work).

Calculations were performed applying the Finite-Element codes SEEP/W (Geostudio, 2004) and PLAXIS (Version 8.2, professional, update-pack 8, build 1499). In the latter case, the exact FE-mesh of the deformation model of Section 4.7.2., page 170 was used. Calculations were performed in each case with the standard settings recommended by the authors of the software packages and applying identical boundary conditions:

Zero flow was allowed through the basal boundary of the geometry model, groundwater heads were prescribed only at its lateral boundaries, with groundwater heights at these points given by the assumed average water level according to Figure 59 (page 134). Models were calibrated adjusting the permeability parameters of the upper landslide layers and the older landslide material via trial and error within the margins and rules given by Table 8 until a good qualitative accordance with this water level was obtained in both simulations. Figure 60 displays the final result obtained with PLAXIS.

Permeability parameters (saturated hydraulic conductivities)

Layer			fixed values and margins m/day	adjusted values m/day	m/s
D	upper landslide layers	D2: vertical	> D1; < 0.09	0.03	3.47E-07
		D1: horizontal	>= C	0.01	1.16E-07
C	main landslide body		0.009		1.04E-07
B	older landslide material		A < B < C	0.004	4.63E-08
A	weathered bedrock		0.0009		1.04E-08

Table 8: Trial-and-error procedure carried out for the calibration of the steady-state Groundwater Flow Calculation.

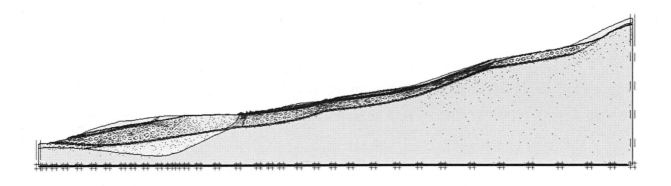

Fig. 60: Result of steady-state Groundwater Flow Calculation with permeability parameters of Table 8; Extreme velocity obtained in the calculated flow field: 2 E-7 m/s.

As only slight adjustments of the parameters were required to attain this goal, with all values remaining in the range of the literature data and within the predefined logical proportions, the hydrogeological assumptions concerning the average pressure level and the average permeabilities of the different layers were found to be of sufficient plausibility.

4.5.5. Generation of two-scenario water levels by means of an Infiltration and Seepage Model

On the basis of the assumptions, which were confirmed by the results obtained in Section 4.5.4.2, two-scenario water levels were generated by simulating both the recharge from a low to a high pore-pressure-scenario and the discharge from a high back to a low pore-pressure-scenario. This was done by means of a 2D slope infiltration and seepage model. As in Section 4.5.4.2, the program SEEP/W was used and the geometry of the model was identical. Infiltration and Seepage Modelling was carried out by *Ronchetti 2007 U.* In the following, a short description of this Infiltration and Seepage Model and its results is given. The applied methodology is described more in detail in *Ronchetti 2007.*

Input flow rate was calculated on a monthly basis using rainfall and snowmelt data recorded at the on-site meteorological station (for location see Fig. 51, page 117) between January 2002 and October 2004. Effective recharge was simulated using a transient flux boundary condition acting on the slope's surface profile. A seepage review function was applied to the slope surface in order to avoid water to pond where there were no ponds in the field. A null-flux condition was imposed at the base of the model and at the upslope model boundary, because the crest of the slope was assumed to represent the phreatic divide. Saturated hydraulic conductivities of the different materials were set as given in Table 8. Unsaturated hydraulic properties were estimated, as first trial, from the value ranges obtained in similar alpine and Apennine earthflows by means of in-situ tests (e.g. *Caris and van Asch 1991, Casagli et al. 2006*). Volumetric water content and conductivity functions were estimated from the grain-size distributions of the landslide materials reported by *Panizza et al. 2006*, following the

137

recommendations within the library integrated in the software package (*Krahn 2004*) and having recourse to functions proposed by other authors (e.g. *Biavati and Simoni 2006*).

Both for the rise from the low pore pressure situation of winter 2001/2002 to the high pore pressures of spring and early summer 2002, as well as for the fall of the water level from the high pore pressure situation in autumn 2002 to the low pore pressures of winter 2002/2003, the infiltration and seepage model was calibrated against the monitoring data of borehole Cpz 2. Water pressure in this borehole – in contrary to borehole C4 – is largely controlled by groundwater regime Type B (see Section 4.5.1., page 133), this means that these measurement series reflect mainly the changes in infiltration which are the object of the simulation. Model calibration was carried out in the form of a trial-and-error procedure by adjusting the functions relating water content and hydraulic conductivity within the margins suggested by the above-cited references. Calibration was continued until a satisfactory fit between calculated and measured groundwater head was obtained at the position of borehole Cpz2, for the recharge as well as for the discharge curve (see Figure 61; for the location of this borehole, see Figure 59 page 134 and Figure 52 page 120).

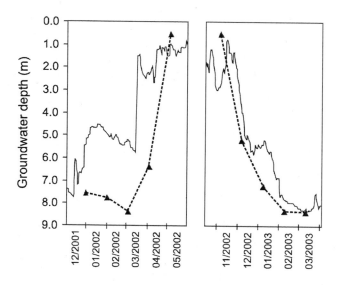

Fig. 61: Observed and calculated groundwater depths in borehole Cpz2 (modified after *Ronchetti 2007 U*).

Since the deformation model does not include partial saturation, the functions relating water content and hydraulic conductivity are not used further, and are therefore not discussed in this work. Hydrostatic pore pressure distributions, starting from the minimum and maximum water level resulting from these calculations are used as low and high pore-pressure scenarios in the deformation model.

4.5.6. Assignment of pore pressure scenarios to different time segments based on climate data

Figure 62 shows how the water-level scenarios were assigned to different time-spans by means of climate data, which were analysed and interpreted together with the measured displacements of landslide source area S3. In doing so, a simplified description of the processes taking place in the field is again required: In the model, for lack of more detailed information, hydrogeological situations as that of rising, falling, low or high pore pressures have to be applied simultaneously to the entire slope; by contrast, the governing climatic factors such as snowmelt, or freezing and thawing of the soil, vary locally with time and also the time-invariant hydrogeological properties of the landslide materials show considerable local variations. For this reason, the climate data of the meteorological station were compared to the average value of the horizontal displacements of GPS-benchmarks 23-28, for these values represent the measurement data of highest quality (see Section 4.2.4., page 122) and in this manner the different local effects are averaged.

When looking at these data, it becomes clear that the displacement rates between July and September are similar for all three years, while being relatively small. On closer examination of the climate data, it appears obvious that these low displacement rates do not correlate with daily rainfalls but with the relatively high temperatures, before and during these time intervals (as postulated already in Section 4.2.4. on page 124). The inclination of the parallel thin grey lines in the upper part of Figure 62 indicates the average displacement rate measured in this time of the year. A similar low average rate results from the displacements between October 2001 and June 2002. With the exception of the last 40 days, which most probably did not yet have an effect, this period was characterised by low amounts of precipitation (see the cumulative curve in Fig. 62) and predominantly by frosty temperatures.

As it is shown by the upper part of Figure 62, the displacement rates of the time spans discussed above must be lower than the average rates of the time intervals in-between, which are not covered by monthly measurements. Therefore, it is assumed that these calm phases are governed by a low pore pressure situation.

With regard to the definition of a representative duration of the transition period representing the delayed reaction of the (deep-seated) slope movements to the climatic conditions, the pronounced change of the infiltration conditions at the beginning of May 2002 is of particular interest: Since early in October (see also above), only small amounts of precipitation were registered and that is why also snowmelt was unincisive.

The drastic turnabout took place on 3rd of May 2002 with a heavy rain event: distributed over three days, 123 mm of rain were recorded (visible in Fig. 62 in the form of a jump of the cumulative curve), and this event was followed by further frequent and partly also abundant rainfalls in the subsequent days and weeks. Until 15th of June 2002, low displacement rates were recorded at benchmarks 23-28, suggesting that this event did not have a significant effect on the slope movements in source area S3 before this date. After short-time acceleration, low displacement rates were again measured starting from 15th of July. Therefore it can be concluded that at this point in time, pore pressures were again falling, or rather, they had already decreased. On account of these observations, transition or delay times which are much longer than 40 days are considered to be unrealistic.

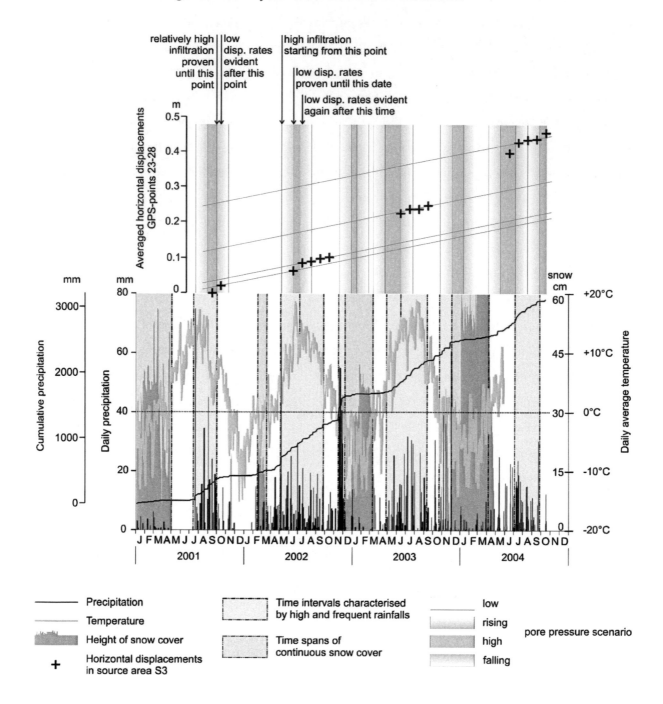

Fig. 62: Assignment of pore pressure scenarios to different time segments based on climate data.
(Meteorological and GPS raw data from *Corsini 2009 U*).

Further indication against a longer transition time is given by the situation in October 2001: Since July, climate was characterised by frequent and partly abundant rainfalls. Cool weather starting from the beginning of September additionally favoured the infiltration process during the last weeks of this wet period. As could be expected, displacements measured between 15th of September and 15th of October were relatively large. (Thanks to the high measuring accuracy of the applied GPS method with respect to horizontal displacements and due to the averaging of 6 values, it is very improbable that this finding is fortuitous).

Time startig from	Assigned pore pressure scenario	Explanatory statement
Middle of July 2001	rising	Onset of frequent and abundant rainfalls
+40 days	high	Very humid climate continued
End of Sept. / Beginning of Oct. 2001	falling	Commencement of dry period
+40 days	low	Dry period continued, long phases with ground frost, unimportant amount of snowmelt, relatively dry spring
Early in May 2002	rising	Beginning of wet phase with high and frequent rainfalls
+40 days	high	Wet weather continued
Early in July 2002	falling	Low displacement rates measured in the field starting from 15th of July; Can not be explained by the rough graphical representation of climate data in this work, showing, by and large, a rainy summer with similar rainfall frequencies and intensities between May and September; Change assumed to have been brought about by dry period from 5th to 14th of July (scarce precipitation: 0.2 mm on 5th, 5.6 mm on 7th and 0.4 mm on 11th) together with summerly temperatures (daily averages definitely above 10°C until early August)
+40 days	low	Rainfall amount and frequency decreasing towards the end of the summer; Followed by dry autumn weather between late September and mid-November
Middle of November 2002	rising	Repeated heavy rainfalls and ephemeral snow
+40 days	high	Delayed reaction to continued wet weather conditions until early December
Middle of January 2003	falling	Snow cover and frost since early December, restrained infiltration for 40 days
+40 days	low	Unchanged winterly conditions
Early in March 2003	rising	Daily average temperatures significantly above 0°C, drastic decrease of snow height
+40 days	high	Snowmelt added to spring precipitation, increasing rainfall amounts starting from April
Middle of June 2003	falling	Low displacement rates measured in the field between July and September; Pronounced rise in temperature reaching summerly values by mid-June
+40 days	low	High temperatures until the end of August; September and most of October relatively dry, except isolated and short phases of heavy rain
Late in October 2003	rising	Repeated rainfalls and ephemeral snow
+40 days	high	Delayed reaction to continued wet weather conditions until the beginning of December
Early in January 2004	falling	Restrained infiltration due to snow cover and frost for circa 40 days
+40 days	low	Unchanged winterly conditions
Early in April 2004	rising	Rapid decrease in snow height
+40 days	high	Melting of thick snow cover (up to 1.15 m at meteorological station at 1700 m.a.s.l.) taking until late April; Moderate but considerable spring rainfalls adding to meltwater; High average displacement rate still measured between mid-June and mid-July
Early in July	falling	Constrained by low displacement rates between 15th July and 15th Sept.
+40 days	rising	Constrained by high displacement rates between 15th Sept. and 15th Oct.; Very frequent rainfalls in July and August

Table 9: Interpretation underlying the assigned durations of the different pore water pressure scenarios.

This time was followed by a long dry period, which, depending on the interpretation of the meteorological data, started somewhere between 26th of September and 5th of October. If the wet period would still have had a significant effect on the movement of the landslide more than 40 days after its end, then the measured average displacement rates between 15th of October 2001 and 15th of June 2002 certainly would have been larger than the low rates typically observed between July and September.

Due to the above considerations, the durations of the transition periods in the simplifying hydrogeological model were uniformly set to 40 days. Taking into account this transition time interval, the entire measurement period between 2001 and 2004 was divided into phases of low, rising, high and falling pore water pressures based on the climate data and the measured average displacement rates. The assignment of these phases is expounded in Table 9 so that in combination with Figure 62 the interpretation of the climate data can be followed in detail by readers interested in this matter.

4.6. Inverse Modelling Concept

In sections 4.6.1. and 4.6.2., the inverse modelling concept applied in this work is described according to *Meier 2008* and *Meier et al. 2008*. Section 4.6.3. describes the strategy developed and followed in the context of this thesis for adapting the numerical models of the forward problems in order to obtain a best possible performance of the inverse modelling methods. In Section 4.6.4., the statistical analyses, which are part of the inverse modelling concept of *Meier et al. 2008*, are explained.

4.6.1. Working scheme of the Inverse Parameter Identification technique

The starting point of the parameter identification technique applied in this study is given by an ordinary geotechnical modelling-task, the so-called forward calculation. This forward problem consists of a specified geometry with given initial and boundary conditions and a material model, which requires a set of material parameters to be determined. It is generally also possible to identify geometrical parameters (*Meier et al. 2006*), but this issue will not be discussed here.

For the first run of the forward calculation, the preset of the parameter vector is done by a random generator within values margins specified by the user. The relevant results of the forward calculation are then read out and their deviation from a set of measured reference data is determined by means of an objective function. This procedure is repeated many times, while at any one time, an optimisation algorithm, based on the parameter combinations and the values of the objective function during the previous forward calculations, identifies an improved parameter set to be used in the next forward calculation.

142

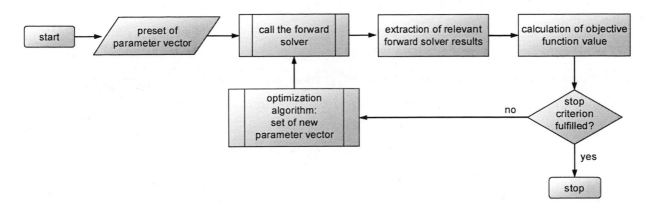

Fig. 63: Flowchart of the iterative procedure (*Meier et al. 2008*).

This sequence of cycles, illustrated in Figure 63, is interrupted when one of the following stop criteria is fulfilled:

(i)   a maximum number of runs or maximum overall calculation time is reached;

(ii) the deviation from the reference dataset, described by the value of the objective function, falls below a specified limit;

(iii) the deviation could not be lowered during a certain number of cycles.

Hence, a direct approach as described by *Cividini et al. 1981* is used here to solve a back-analysis problem. In an iterative procedure, the trial values of the unknown parameters are corrected by minimising an error function. It is therefore not necessary to formulate the inverse problem itself, the desired solution is obtained by combining the results of numerous forward calculations with an optimisation routine. To quantify the deviation between the reference data and the modelling results, the frequently used and relatively simple method of least squares was chosen. In this method, the objective function f (x) for more than one reference dataset is defined as:

$$(1) \quad f(x) = \sum_{g} [w_g \cdot f_g{}'(x)]$$

$$(2) \quad f_g{}'(x) = \frac{1}{m} \sum_{h=1}^{m} w_h \left(y_g^{\,calc}(x) - y_g^{\,meas} \right)_h^2$$

In (1) and (2), x denotes the parameter vector to be estimated and w_g are positive weighting factors associated correspondingly with the error measure $f_g{}'$ (x). Via the weights w_g, the different series g can be scaled to the same value range and different precisions can be merged, for example, a series of measuring data possessing a higher precision is included with a higher weighting factor compared to more uncertain data. The weighting factors w_h are used to provide a possibility for considering different precisions and measurement errors within one and the same data series.

For minimising the objective function, i.e. for finding its global minimum inside the n-dimensional search space spanned by n axes representing n model parameters to be identified, different optimisation algorithms can be used. In this work a Particle Swarm Optimisation algorithm (PSO), whose efficiency had already been proven for more simple examples from the Corvara case study (*Meier et al. 2008*), and a Shuffled Complex Evolutionary algorithm (SCE) were applied.

4.6.2. Adaptation of the Inverse Modelling Concept for geotechnical applications and short description of the algorithms applied in this work

The parameter identification strategy described above is common in many engineering fields, but it had to be tested and adapted especially for geotechnical applications (*Meier 2008*). Geotechnical problems are often characterised by highly nonlinear system responses and a rough objective function topology, sometimes with saltuses. In many cases, there are large uncertainties concerning the parameter values, consequently failed forward calculations occur and there are restrictions of the parameters and dependencies between them. As an example, the parameter Cc (Compression Index) of a given soil cannot be smaller than its Cr (Recompression Index), and if the stiffness of this soil increases, also its friction angle will rise. Frequently, complex model geometries are combined with quite sophisticated constitutive equations, resulting in long calculation times of the forward problem. As a consequence, any technique that would require an initial guess of the parameter set was discarded, and, as a matter of principle, the overall search procedure was designed in a way that the user just defines the upper and lower margins of a search interval for each of the parameters. In order to cope with the above-described characteristics, only selected optimisation algorithms are used. Furthermore, these algorithms were modified in some small details with respect to their application to this class of problems (*Meier 2008*). For example, in the case of the PSO and SCE algorithms which are applied in this study, randomly generated additional supporting points help reviving the search in other areas of the search space, if the parameter vector gets "trapped" in the surroundings of a local minimum of the objective function or if it moves towards a "forbidden zone" marked by physical or technical constraints. This is done to avoid too many futile calls of the forward calculation with parameter sets from these unpromising regions. Usually, with respect to the calculation time of the forward problem, the time consumption of the inverse modelling procedure itself plays a minor role, this is why complex but powerful optimisation algorithms are advantageous.

For the same reason, when adapting the Inverse Parameter Identification Strategy for geotechnical applications, the main focus was on keeping the number of required forward calculations as low as possible. Therefore the optimisation routine was linked to an "intelligent" database, which records and

144

controls the calls of the forward solver (in this work: the Finite-Element program). By this means, repeated forward calculations with identical parameter sets can be prevented. The database also allows for imposing rules, e.g. constraints, on the setting of the parameter vector, i.e. on the way different parameter values can be combined within one parameter set. These rules may be defined based on technical or physical considerations with respect to the studied problem. In this manner, unfavourable parameter combinations can be filtered out already before a definitely unnecessary or failing forward calculation would be started. In the following, the optimisation methods used in this work will be described shortly and qualitatively (detailed, also mathematical, specifications are given by *Meier 2008*).

Particle Swarm Optimisation Algorithm (PSO)

The Particle Swarm Optimisation Method forms part of a group of algorithms called Population Based Methods, whose basic working principles derive from the behaviour patterns of individuals in living nature, e.g. fish shoals or swarms of bees. The task of searching the objective function topology for its global minimum is performed by a (mathematical) set of coexisting individuals, which interact according to a system of rules of conduct. Particle Swarms, as they are used in this work were originally developed by *Eberhart and Kennedy 1995* (see also *Kennedy and Eberhart 1995*). This method has its roots in Artificial Intelligence and Social Psychology. It can be seen as a mathematical reproduction of a swarm consisting of a finite number of punctiform particles. Each of these individuals is assigned a position within the solution space and a velocity vector, which, via its direction and norm, defines the movement of that particle. The particles are moving without colliding through the solution space, searching for the global minimum of the objective function. In doing so, each one of the particles is able to update its movement vector after each calculation cycle, based partly on its own movement history and partly on data communicated to it by other particles. Furthermore, due to the implementation of a so-called "Particle Craziness", following suggestions of the above cited authors, in some cases also illogical changes of particle movement are executed. This feature helps bringing the search out of apparent local minima.

Shuffled Complex Evolution Algorithm (SCE)

The Shuffled Complex Evolution Method belongs to a group of algorithms usually referred to as "Combining Methods". They were developed by combining the methods and strategies of different optimisation algorithms in order to overcome weak points, restrictions and disadvantages of the individual methods. The SCE algorithm used in this study was proposed by *Duan et al. 1993*. It comprises elements of Stochastic Optimisation Methods, of Simplex Based and Complex Based Methods, and of Evolutionary Methods. Stochastic Methods are algorithms which are following combinatorial or random approaches. When applying Simplex Based Methods, the objective function topology is searched by scanning it with the help of a Simplex, i.e. a (mathematical) point set that geometrically represents an n-dimensional polytope determined by n+1 parameter vectors, whereas Complex Based Methods are making use of sets of simplexes. Evolutionary Methods imitate basic principles of the evolution of natural species, i.e. recombination, mutation and selection. The SCE algorithm combines these different methods by means of a sophisticated system of inner and outer loops. In so doing, a very efficient global search strategy is created, which is able to manage even very rough objective function topologies that may show defects and mathematical discontinuities or may be segregated into different search regions.

4.6.3. Adaption of the numerical model of the forward problem for its use in the optimisation routine

While the Inverse Modelling concept concentrates on lowering the number of necessary forward calculations, the numerical model of the forward problems was adapted in order to minimise the runtime of the single forward calculations and thus to allow a high number of calls. For this purpose, the numerical model has to be reduced and/or coarsened. In contrast, the forward model must still be sufficiently detailed to reproduce the reference data with the required accuracy. To negotiate this dilemma, the forward models were built in two versions: A full-scale model, designed to obtain a high accuracy of the simulation results and a reduced model version designed to be used in the optimisation routine. For building this model, a copy of the full-scale model was modified in a way that it allows for a much faster forward calculation but results still do not differ substantially from those produced by the full-scale model. Whether two results can be termed "nearly identical" has to be decided considering the general level of accuracy that can actually be reached by the simulation of a geotechnical or engineering-geological problem, which is affected mainly by the heterogeneity of the materials and by the inexactness of the model assumptions. Therefore, this judgement ultimately depends on the opinion of the observer. For this reason, in Section 4.7., (starting from page 148) the final results of both model versions will always be displayed together and in comparison to the reference data, so that the reader gets an idea of the difference. As a next step, it is essential to select the parameters to be identified and to decide on the upper and lower limits of their plausible values margins. The values of some of the parameters might be fixed with the aid of previous knowledge. By this means, the search space can be narrowed down. These specifications require the experience of the geoscientist and must be done with care, since they influence the obtained results.

4.6.4. Statistical analyses

Due to the inhomogeneities of geological materials involved and the uncertainties related to the initial conditions and the geometrical boundary conditions, geotechnical problems tend to be underdetermined. In order to ensure and improve the efficiency of the back-analysis, it is of significant importance to check if the set of parameters to be identified may be reduced and if for the prescribed trusted zone the optimization problem is well posed. For this purpose, a statistical analysis is done based on the results of a Monte Carlo procedure including a sufficient parameter set. Figure 64 shows the principle scheme of the matrix plot used here to visualise the results of the Monte Carlo simulation. A standard mathematical tool for examining multidimensional datasets is the scatter plot matrix (*Manly 1944*), which is included in the matrix plot presented in Figure 64, where each nondiagonal element shows the scatter plots of the respective parameters.

The matrix is symmetric. Matrix element D-B, for example, may suggest that the involved parameters B and D are not independent, but strongly correlated. The diagonal matrix plot elements (A-A, . . . , D-D) show plots where the value of the objective function is given over the parameter which is associated with the corresponding column. The lower the position of a data point in this plot, the smaller is the corresponding objective function value. These plots are called hereafter "objective function projections".

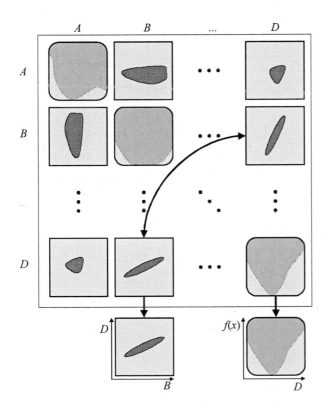

Fig. 64: Scheme of the scatter plot matrix (from *Meier et al. 2008*).

147

If the problem is well posed, i.e. well determined, each of these plots of the objective function projections has to present one firm extreme value as it is the case in the diagram D-D. Otherwise, the respective parameter could probably not be identified reliably. By filtering out data points that have objective function values larger than a certain threshold level, the distribution of the remaining points gives a rough idea of the size and shape of the extreme value space (solution space). For further statistical analyses, the well-known linear 2D correlation coefficient can be calculated from the individual scatter plots. Referring to *Will et al. 2003,* variables with a correlation coefficient of less than 0.5 are considered as "noncorrelated".

4.7. Deformation Modelling

Based on the field data and model assumptions described in the previous Sections, the deformations of the slope and its (geological) materials were calculated applying the Finite-Element method. All these calculations were carried out using the commercial code PLAXIS (Version 8.2, professional, update-pack 8, build 1499) and taking into account the effect of large deformations by means of an updated Lagrangian formulation (updated mesh analysis) (*Reference Manual PLAXIS V8 2003*). The Calibration of the numerical models was performed by identifying constitutive parameters of the shear-zone materials via/applying the Inverse Modelling concept described in Section 4.6.

4.7.1. Simulation of laboratory tests

For the deformation rates calculated by the numerical slope model, the material parameters of the shear zones are essential. The properties of these zones are governed primarily by the clayey soil matrix of the landslide material (see Section 4.3.3., page 129 and Section 4.4., page 131). Therefore simulations of laboratory tests are carried out to verify if the constitutive model chosen, in combination with the adopted calculation method and parameter identification technique, are able to realistically reproduce the deformation behaviour of such a material, for the well-known dimensions and precisely defined stress ratios and stress paths of the different experiments. Only if this verification is successful, the same approach can be used for calculating the deformations of the slope and for calibrating this numerical slope model against the displacement measurements, regarding slope deformation as an experiment whose boundary conditions are only partly known.

The other parts had to be defined based on engineering-geological assumptions which always – even in the case of the extensively investigated site of Corvara – remain uncertain to some degree. Whether the calibrated numerical model is realistic therefore depends to a large extent on the quality of these engineering geological assumptions.

148

In order to obtain representative soil material for the laboratory tests, it was originally intended to take samples from the borehole cores of C4 and Cpz2 at that depth where the inclinometer measurements indicated the shear-zone (see Fig. 52, page 120 for the location of these boreholes). However, the quality and heterogeneity of this material did not allow for manufacturing a sufficient amount of reproducible specimens needed to ensure the comparability of different tests. Anyway it was questionable if the punctual picture of material type and properties encountered in these two neighbouring cores is automatically the only one representative for all shear-zones, altogether extending over a length of more than 1000 metres (when considering only their 2D profile). Therefore the borehole cores, also these from other landslide sectors, were classified with respect to their stratigraphical composition and soil mechanical properties (compare Chapter 2 pages 26 to 37, and the Appendix part A), examining them at shear zone depth and also in between. Furthermore, also the landslide materials exposed in many outcrops and in areas bare of vegetation were taken into account. Based on this classification, clay-rich soils composed of completely disintegrated weathering detritus, mainly originating from the La Valle formation and partly from the San Cassiano formation, were found to be most representative. Table 10 shows some characteristic soil mechanical parameters of this type of material, giving their value ranges together with the number of classified samples. In Table 10, only the data of samples which are most similar to those used in the laboratory tests that are described and simulated in the following Sections are shown. Unless otherwise expressly described in the text, all classifications and tests described in the following were carried out according to the German standard DIN. A sufficient amount of material of this sort was then sampled circa 50 cm below the ground surface in natural outcrops generated by the incision of small streams and by superficial sliding of the vegetation cover and correspondence of these samples with the above-described characteristics was tested in the laboratory.

Percentage of clay particles (diameter < 2 µm)	26 - 34 % (4)
Density of grains (g/cm³)	2.76 - 2.73 (4)
Lime content	32 - 36% (4)
Loss of ignition	4.2 - 6.6 % (5)
Liquid Limit w_L	0.50 - 0.65 (5)
Plastic Limit w_P	0.25 - 0.29 (5)

Table 10: Characteristic properties of the studied soil (number of classified samples in parentheses).

The specimens for two oedometer tests (1a and 1e) were both made from sample 4.120 (see Fig. 56, page 128 for the location of the sampling point) which was taken with a soil sample ring. From the interior of this cylinder (15 cm in diameter and 8 cm high), the much smaller (7 cm and 2 cm, respectively) oedometer samples were later cut out, trying to capture only the matrix fraction without rock fragments > 2 mm (after the test, the specimens were destroyed and passed through a sieve to make sure that no larger stones had affected the result). This procedure was followed to preserve density and water content of the fresh sample under field conditions. Of course, these conditions are not those prevailing at depth along the shear-zones, but it was preferred to use these values instead of choosing initial density and water content for the laboratory tests arbitrarily.

As the samples for the tests in the triaxial apparatus (5 cm in diameter and 10 cm high) were longer than the oedometer samples, it turned out very difficult to cut them from a larger soil body without including coarser components (> 2 mm). For this reason, a mixed sample was prepared from sample materials of similar stratigraphic composition (as sample 4.120) taken at two different locations next to the sampling point of sample 4.120. From this sample, the fraction < 2 mm was separated by sieving and then used to manufacture specimens with similar water content and density. Table 11 shows these values for the three different samples used in the laboratory tests described in this study.

	Sample 4.120_oed_1a	Sample 4.120_oed_1e	Prepared sample 6.197
Water content	0.43	0.41	0.41
Bulk density (g / cm³)	1.79	1.87	1.82
Dry density (g / cm³)	1.25	1.32	1.30
Porosity n	0.54	0.52	0.53

Table 11: Water content, dry density and porosity of the test specimens.

4.7.1.1. First Oedometer test: stress-dependency of the soil stiffness

The first Oedometer test was performed in order to investigate the stress-dependency of the soil stiffness and the course of the load-dependent settlements with time including consolidation and creep behaviour. The stress-range of interest was defined according to the overburden pressures present along the basal and secondary shear-zones (circa 100-800 kPa, compare Section 4.3.2., page 127). The test was conducted in a fixed Oedometer ring with an inner diameter of 71.45 mm and a height of 20.21 mm. Drainage was allowed on the top and at the bottom of the soil sample.

First, the sample was preloaded with 9 kPa during two days and then with 13 kPa during one day. Two hours after the application of the 9 kPa load step, the case containing sample, oedometer ring and filter plates was filled with water to prevent drying of the soil during the test. The reference data set starts at the end of preloading. Then the load was doubled successively, with each load step lasting 24 hours, loading the sample with 25, 50, 100, 200, 400, and 800 kPa. After that, it was unloaded at 400, 200, 100, and 50 kPa, and finally it was reloaded again with 100, 200, 400, and 800 kPa (last step took 43 hours). The displacements of the sample top were recorded continuously. Reference data points were set at the beginning and at the end of each load step and, within the duration of the respective load step, after 10, 20, 40, 100, 200, 400 and 1000 minutes (calculated from its start).

For the numerical model of the test, an axisymmetric geometrical configuration with the exact dimensions of the specimen was used. Horizontal fixities were assigned to the lateral boundary and to the rotation axis, simulating the stiff oedometer ring, and vertical fixities were attributed to the basal boundary, representing the fix filter plate at the bottom. The full-scale version of the model was discretised with 1206 triangular 15-node elements. For the model later used in the optimisation routine, the number of these elements was reduced to 30, which was found to be the fastest possible configuration still producing acceptable results. In the simulation, after generating the initial stress state by applying the soil self-weight (gravity loading procedure), distributed loads were applied perpendicular to the top boundary analogue to the laboratory conditions.

A first trial forward calculation was carried out using parameter values determined experimentally. Three parameters of the Soft Soil Creep Model can be derived directly ($\lambda^*$and μ^*), or determined approximately (κ^*) from the oedometer test (*Vermeer and Neher 1999*). Shear test data of similar soil samples from the Corvara site, reported by *Panizza et al. 2006* is shown in Table 12. These values were averaged, giving double weight to the triaxial test data, which were assumed to be more precise, arriving at an average friction angle of 20° and an average cohesion of 27 kPa.

	effective friction angle φ' (°)	effective cohesion c' (kPa)
direct shear tests	18	20
	18	10
	20	49
	18	39
	16	49
	14	69
	18	25
	20	20
triaxial shear tests	19	7
	28	14

Table 12: Shear test data from the landslide material at Corvara (data from *Panizza et al. 2006*).

Due to extensive alteration by weathering processes and disturbance by movements, the clayey soil matrix of the landslide material is presumed to be in a near-normally consolidated state almost everywhere. Figure 65 shows the application of Janbu's concept for the assessment of preconsolidation stress (*Janbu 1967*) to the soil sample of Oedometer test 1a. The approximately linear increase of the Tangent Modulus M suggests that over the whole first-loading stress range of the experiment, the soil sample may be considered as normally consolidated. Along the shear-zones of the slope, the clayey soil matrix has undergone long-lasting shear deformations and is therefore assumed to have reached a residual state (compare Section 4.4, page 131). Consequently, volume change due to further shearing is considered to be negligible. For these reasons, the dilatancy angle was set to zero in all calculations of this study.

A permeability test conducted in the triaxial apparatus using an equivalent soil sample (number 4.121, see pages A129 and A130 in the Appendix) gave a coefficient of water permeability of $2*10^{-10}$ m/s. The set of experimental parameter values is given in Table 14 (page 156) and in Figure 68 (on page 156) the results of the forward calculation, done with the full-scale numerical model and using these parameters, are displayed together with the measured reference data of the Oedometer test. The graph shows that qualitatively the course of the deformations with time is expressed quite well, but the deformations are underestimated by the simulation.

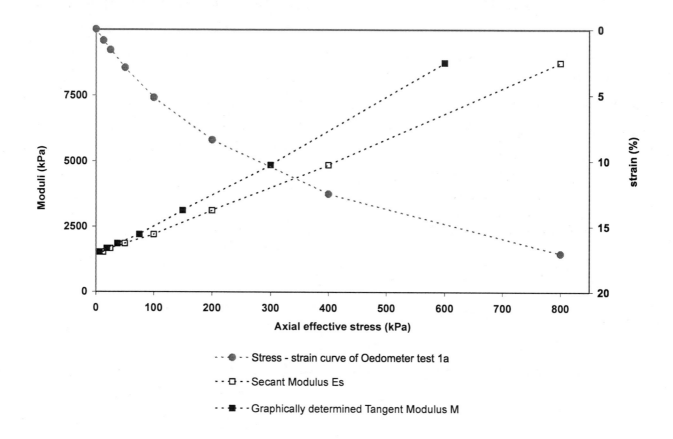

Fig. 65: Assessment of preconsolidation stress for sample 4.120.

In order to increase the determinateness of the inverse problem and to reduce the number of parameters to be identified at the same time, the process of model calibration was split in two steps. In a first step, the laboratory test was simulated by means of an elastic-plastic deformation analysis without considering the time-dependent development of pore pressures, i.e. applying the load steps under drained conditions, and the final deformations observed at the oedometer at the end of the respective load steps were used as reference data for model calibration. In this step, a maximum possible number of parameters were identified. Then, in a second step, the complete course of deformations with time was simulated by means of a consolidation analysis, and the rest of parameters were identified.

4.7.1.1.1. Elastic-plastic deformation analysis

Statistical analysis

For investigating model behaviour and for deciding on the parameters to be identified in the first step, a statistical analysis comprising 2000 calls of the forward calculation (compare Section 4.6.4., page 147) was carried out. Thereby all material parameters, except the dilatancy angle, were varied within wide search areas. These search intervals are shown by Table 13.

parameter	fixed parameters	varied parameters			
		minimum	maximum	ln(min.)	ln (max.)
φ (°)	20	(8)	(30)		
ψ (°)	0				
μ^*	1,46E-03	(0,00001)	(0,75)	(-11,51)	(-0,29)
c (kPa)		0,001	100		
κ^*		0,001	0,5	-6,91	-0,69
λ^*		0,002	1	-6,21	0

Table 13: Search intervals for the parameters of Oedometer test 1a. (Parameter limits in parentheses were not used in the first step of the optimisation procedure).

For the parameters λ^*, κ^* and μ^*, logarithmised values had to be used for defining these intervals, in order to avoid an overrepresentation of high parameter values during the search. Furthermore, a parameter constraint (s. Section 4.6.2., page 145) was prescribed, demanding for $\lambda^* > \kappa^*$, because this has to apply for all materials. The scatter-plot-matrix presented in Figure 66 displays the parameter values of all calculations with objective function values lower than 10^{-7}. The modified compression index λ^* and the cohesion c appear to be correlated (correlation coefficient of 0.88), to a lesser extent, this holds true also for λ^* and κ^* (correlation coefficient of 0.65) and for κ^* and c (correlation coefficient of 0.64). The objective function projections of c, κ^* and λ^* indicate that data points belonging to good model fits seem to concentrate in quantifiable value ranges of these parameters. Therefore, all three of them could be selected for identification via the optimisation procedure. In contrast, the simulation is apparently neither sensitive to φ nor to μ^*. As in the Oedometer, shear failure is prevented by the virtually non-deformable oedometer ring, model behaviour can be considered realistic with respect to the first point. In this simulation, only the deformations at the end of the load steps are calculated. As creep deformations account for a relatively small part of these deformations, the statistical analysis provides no clear information on the range of the best-fit values of the modified creep index μ*. By its shape, the objective function plot only gives a vague hint that possibly the values with the smallest deviations from the reference data might be obtained in the lower part of the search interval, but not until its lower boundary. After all, this is roughly where also the experimental value lies. According to the findings of the statistical analysis, for the back-analysis, the friction angle φ and the modified creep index μ* were fixed on their experimental values.

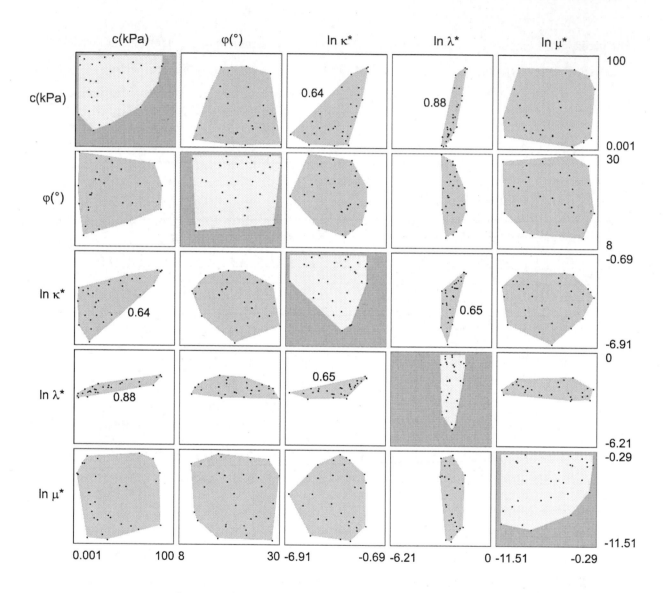

Fig. 66: Scatter plot matrix for Oedometer test 1a (modified after *Meier 2009 U*).

Optimisation with two different algorithms

After the statistical analysis, the Particle Swarm Optimisation Algorithm (PSO) and the Shuffled Complex Evolution Algorithm (SCE) described in Section 4.6.2. (page 145) were run to calibrate the model and to identify the three parameters (c, κ* and λ*) within the search intervals of Table 13. After

1590 calls, the PSO algorithm had managed to reduce the deviation to $5*10^{-9}$, which was regarded as a sufficiently low value to stop the optimisation routine. The SCE reached a slightly better fit (deviation value of $3*10^{-9}$) already after 557 calls of the forward calculation, i.e. three times faster. As from then on, no lower objective function value was found during several further cycles, this optimisation was stopped, too. Both identified parameter sets are given in Table 14 (page 156) and the corresponding calculation results can be seen in Figure 67.

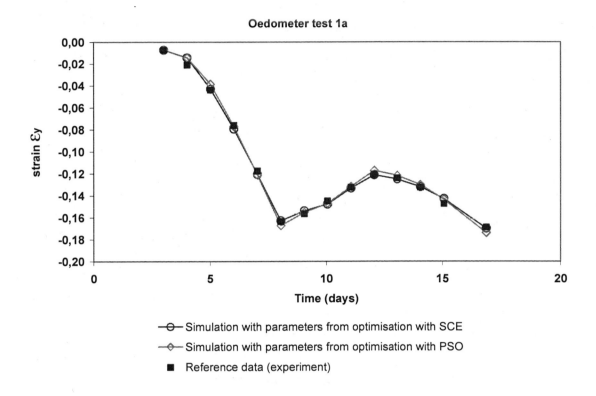

Fig. 67: Calculation results for Oedometer test 1a, first step of model calibration. Deformations at the end of the load steps. (Laboratory test and simulation: this study; optimisation: *Meier 2009 U*).

4.7.1.1.2. Consolidation analysis

In the second step of model calibration, the parameters c, κ^* and λ^* were fixed at the values that were identified (by the SCE algorithm) as those producing the best possible fit between the simulation and the reference data. The friction angle was set to its experimental value while μ^* was varied within the margins already used in the statistical analysis (see Table 13, page 153).

The simulation of the detailed course of deformations with time demands the specification of the permeability of the soil, i.e. of at least one additional parameter. At the sample scale, the matrix of the landslide material exhibits neither any preferred orientation of its components nor layering, therefore, only an isotropic coefficient of water permeability, termed "k", was used. The search interval for this parameter was defined between 10^{-13} m/s and 10^{-8} m/s. Logarithmised values were used for defining

155

the search area to avoid an overrepresentation of high parameter values during the search. For calibrating the model and to identify μ* and k, the SCE algorithm was applied. After 496 calls, the deviation could not be lowered significantly any more. The best parameter set is shown in Table 14 and Figure 68 depicts the results of the simulation obtained with this set for both the model version used in the optimisation routine and the full-scale model. When looking at these data, it becomes clear that the exact values of the identified parameters are mesh-dependent, i.e. slightly different parameter values would be needed for an identical reproduction of the best fit with the full-scale model.

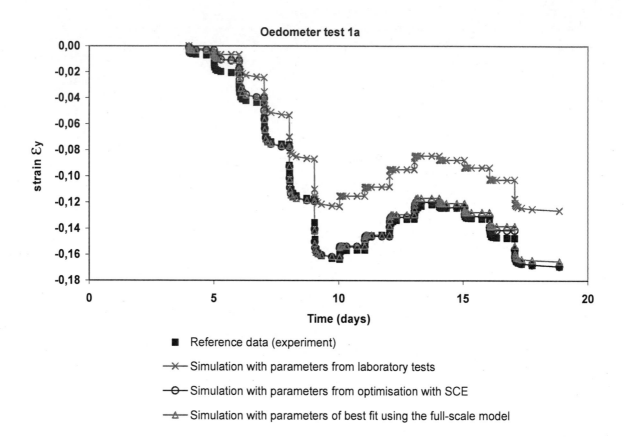

Fig. 68: Results of consolidation analysis (second step of model calibration) for Oedometer test 1a.

(Laboratory test and simulation: this study; optimisation: *Meier 2009 U*).

Parameter	Experimental values	Identified values		2nd step
		1st step		
		PSO	SCE	SCE
φ (°)	20			
ψ (°)	0			
c (kPa)	27	25,1	20,2	
λ^*	0,064	0,082	0,075	
κ^*	0,035	0,051	0,038	
μ^*	1,46E-03			1,31E-03
k (m/s)	2E-10			8,3E-11

Table 14: Summary of parameter values for Oedometer test 1a.

(Laboratory test and simulation: this study; optimisation: *Meier 2009 U*).

156

All in all, the course of the deformation of the sample is simulated with good accuracy: except for the first two phases of the simulation (load steps 25 kPa and 50 kPa) no serious under- or overestimation is observed. But it can be noticed that the quality of reproduction of the reference data varies for the different load steps. This is because in contrast to the simulation, both μ^* and k are not constant over the whole experiment but vary with the stress range and the resulting compaction degree of the soil. Further parameters would be needed to describe these changes. But this would decrease the determinateness of the problem. In the inverse model, this is expressed by an expansion of the search space into an additional dimension for each additional parameter. Therefore, considering the slope of Corvara, where already the values of μ^* and k, themselves, are marked by much larger uncertainties than in a laboratory test, the attempt of introducing and identifying additional parameters related to them is very unlikely to make the model more realistic.

4.7.1.2. Second Oedometer test: prolonged creep phases

A second oedometer test was performed to study deformation behaviour during prolonged creep phases in the medium stress range (with respect to the overburden pressures of the shear-zones). At its beginning, the experiment was conducted in the same manner as oedometer test 1a, but after first-loading from 200 kPa to 400 kPa, the soil was left creeping during 10 days. Then it was unloaded back to 200 kPa and again left creeping during 6 days before the test was stopped.

The numerical model of this test was built analogue to that of oedometer test 1a, the full-scale version was discretised with 1226 triangular 15-node elements and the reduced version with 32 elements of the same type. Again, a first-trial calculation was carried out with experimental parameter values. This parameter set is shown in Table 16 (page 160). Deriving the parameters λ^*, κ^* and μ^* from the test resulted in slightly different values, due to the natural heterogeneity of the material and due to the lower stress-range under investigation. The results of this simulation (obtained with the full-scale model) are presented in Figure 71 (page 160), together with the reference data. Also for this test, the deformations are underestimated by the simulation, particularly at the first load-steps, which represent the low stress-range. Calibration via back-analysis was split in two steps, in the same way as for the first oedometer test.

Statistical Analysis

Based on the considerations of Section 4.7.1.1.1 (page 153), a statistical analysis with 1375 forward calculations was now performed only for the parameters c, κ^* and λ^*. Figure 69 displays a scatter-plot matrix with the parameter sets of all 73 simulation-runs that gave back objective function values smaller than 10^{-8}. The matrix is similar to that of oedometer test 1a, the parameters λ^* and c are again correlated (correlation coefficient 0.91), and a less pronounced correlation could also be found for κ^* and c (correlation coefficient 0.66).

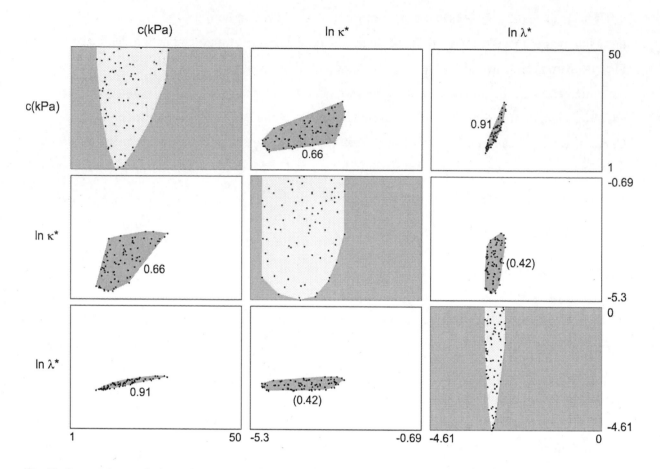

Fig. 69: Scatter plot matrix for oedometer test 1e (modified after *Meier 2009 U*).

As a consequence of the reduction of the number of varied parameters and also of the narrower search intervals, the search space has shrunk and the determinateness of the problem has increased. This is expressed by the more peaked objective function projections of c and λ^* in this matrix compared to the matrix of Figure 66 (page 154). Also the objective function projection of κ^* very clearly points to a minimum region but, by its broader shape, it indicates that the result of the simulation is less sensitive on this parameter than on the other two. This is because the simulated test comprised only one unloading phase, whereas the other five phases are first-loading phases, and in the constitutive model, κ^* is used to describe the unloading and reloading behaviour

Optimisation using PSO and SCE

For the elastic-plastic simulation (first step of model calibration), the results of two optimisations, using the PSO and the SCE algorithm, are presented in Figure 70 together with the final deformations measured in the oedometer test at the different load steps. The upper and lower values margins of the parameters varied are given in Table 15 and the identified parameter sets are presented in Table 16.

parameter	fixed parameters	varied parameters			
		minimum	maximum	ln(min.)	ln (max.)
φ (°)	20				
ψ (°)	0				
μ^*	8.90E-04				
c (kPa)		1	50		
κ^*		0.005	0.5	-5.3	-0.69
λ^*		0.01	1	-4.6	0

Table 15: Search intervals for the parameters of Oedometer test 1e.

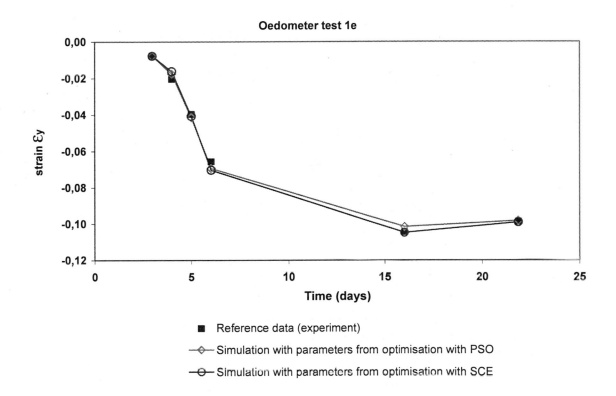

Fig. 70: Calculation results for Oedometer 1e, first step of model calibration. Deformations at the end of the load steps (Laboratory test and simulation: this study; optimisation: *Meier 2009 U*).

The simulation results for the consolidation analysis of the second step, obtained with the finally identified parameter set of the best fit are illustrated in Figure 71 and the corresponding values of μ^* and k, which were found by the SCE algorithm, are also given in Table 16. When looking closely at Figure 71, it can be observed that for the reasons explained in Section 4.7.1.1. (on page 157), relative deformations are slightly overestimated at the fourth load-step (100-200 kPa) and slightly underestimated at the fifth load step (200-400 kPa). Like for the first oedometer test, the identified parameters are similar to those determined directly from the laboratory test. Material behaviour is again described correctly, except for small differences resulting from the simplifications made in the model (e.g. constant values of k and μ^*).

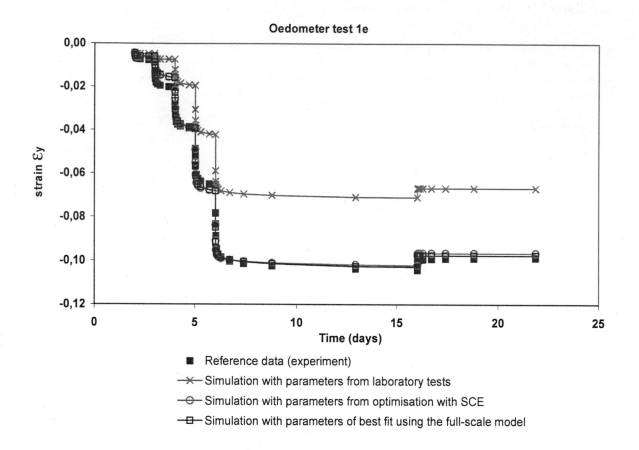

Fig. 71: Results of consolidation analysis (second step of model calibration) for Oedometer 1e. (Laboratory test and simulation: this study; optimisation: *Meier 2009 U*).

Parameter	Experimental values	Identified values		
		1st step		**2nd step**
		PSO	SCE	SCE
φ (°)	20			
ψ (°)	0			
c (kPa)	27	13,4	16,1	
λ^*	0,050	0,052	0,056	
κ^*	0,020	0,014	0,024	
μ^*	8,9E-04			1,33E-03
k (m/s)	2E-10			1,5E-10

Table 16: Summary of parameter values for Oedometer test 1e. (Laboratory test and simulation: this study; optimisation: *Meier 2009 U*).

160

4.7.1.3. Isotropic compression test: consolidation and creep at an isotropic stress state

In order to study consolidation and creep of the material at an isotropic stress state, an isotropic compression test was carried out. The test was performed in the triaxial apparatus on a cylindrical soil sample with a diameter of 5 cm (49.88 mm) and a height of 10 cm (98.55 mm). Drainage was allowed on top and at the bottom of the specimen. All load steps were performed isotropically, applying a hydrostatic cell pressure. The sample was preloaded at 30 kPa for three hours and then at 50 kPa for 17 hours. The reference data set starts at the end of this phase. The sample was then gradually loaded to 800 kPa within one hour, increasing the load by steps of 100 kPa. After reaching the target load, the stress level was left constant for 3 weeks. Top displacements and volume change of the sample were recorded during the whole test. The displacements measured at the sample top were taken as reference data for the simulation. The first reference data point was set at the beginning of the 800 kPa load step, and further reference values were taken 110 minutes, 8 hours, 1 day, 2, 5, 7, 9, 12, 15 and 22 days after this point.

For the numerical model of the test, an axisymmetric geometrical configuration with the exact dimensions of the test specimen was used. The full-scale version of the model was discretised with 1422 triangular 15-node elements. For the model later used in the optimisation routine, still 335 of these elements were needed to ensure sufficiently accurate results. Horizontal fixities were assigned to the rotation axis. Vertical fixities were attributed to the basal boundary, representing the fix filter plate at the bottom. After generating the initial stress state by applying the soil self-weight (gravity loading procedure), two identical distributed loads were applied, one perpendicular to the upper boundary (vertically) and the other one perpendicular to the lateral boundary (radially). Loading was carried out in the same way as in the laboratory, but instead of the stepwise application of the 800 kPa target load, this load was applied in one calculation phase, continuously increasing it in such a manner that the integral of the load as a function of time equals the test conditions.

The results of a forward calculation using the averages of the laboratory values of Table 14 (page 156) and Table 16 (page 160) are shown in Figure 73 (page 164) in comparison to the reference data. At the beginning of the test, the simulation overestimates the deformations, whereas in the following, until the end, they are underestimated.

Statistical analysis

The results of the isotropic compression test are known to be sensitive to the compressibility of the material during first-loading, the ability of excess pore pressures to dissipate, and to the amount of creep deformations. Therefore, the modified compression index λ^*, the coefficient of water permeability k, and the modified creep index μ^* were selected for statistical analysis. As the test contains no unloading or reloading phases, the modified swelling index κ* can not be identified. Therefore, by means of the database, its value was linked to the value of λ*, multiplying this parameter by 0.5, which is the typical λ*/ κ* - ratio observed in a series of oedometer tests performed on this material and similar soils from other landslides of the earth-flow type (*Schädler et al. 2006 U*). The shear strength parameters φ and c were fixed at their experimental values.

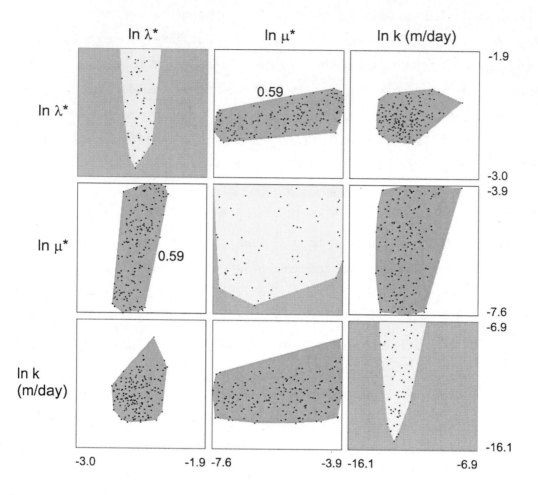

Fig. 72: Scatter plot matrix for the isotropic compression test (modified after *Meier 2009 U*).

The scatter-plot matrix of the statistical analysis is presented by Figure 72. The Monte Carlo procedure involved 1690 calls of the forward calculation. Only the 60 parameter sets related to the lowest objective function values ($< 4*10^{-7}$) are displayed in the matrix plot. The plots of λ* and μ* suggest that these parameters may be correlated to each other (correlation coefficient 0.59). The close relationship between the stress-dependent compressibility of different soil types and the amount of creep deformations they exhibit during secondary compression is an established fact in soil mechanics (e.g. *Mesri and Godlewski 1977, Mesri and Castro 1987*) and is therefore part of the Soft Soil Creep Model.

The objective function projections of λ* and k have sharp peaks whereas that of μ* gives no clear indication on the location of the minimum of the objective function. However, none of the eight best parameter sets has a value of μ* lying in the right half of the search interval. The probability that this finding is purely fortuitous is very low (around 0.5[8]); it is very likely that the peak of this objective function projection will become sharper if more random parameter sets are tested and that a more clear minimum region can be found by the optimisation algorithms. From the plot of the reference data in Figure 73, the lower sensitivity of the objective function value to μ^* can easily be understood: λ^* and k have a very strong influence, respectively, on the total amount of deformations observed in the simulation and on primary consolidation, and therefore on the curvature of the arc described by the data points. In contrast, fairly good model fits can still be obtained for relatively bad estimates of the development of creep deformations with time (last part of the reference points), as long as they are not overestimated in an extreme manner.

Of course, μ^* could be identified with less effort, when separately comparing the simulation results with a reference data set containing only the relative deformations in the last part of the test, e.g. after 7 days. But with respect to the slope model it was preferred not to reduce the complexity of this problem and to check if it is possible to identify the above-mentioned three parameters at the same time. Naturally, in the case of the displacement measurements at Corvara, no straightforward "assignment" of a certain part of a curve to one specific model parameter is possible.

Optimisation using PSO and SCE

Optimisation was then performed by means of the two different algorithms, using the search intervals and parameters of the statistical analysis. An overview on the search intervals is given in Table 17. After 1000 calls of the forward calculation, the PSO algorithm had reduced the deviation to $2.7*10^{-8}$. With only slightly higher calculation effort (1099 calls), the SCE algorithm reached a lower deviation value of $2.0*10^{-8}$. The higher quality of the corresponding model calibration is mainly due to a better reproduction of the creep rate. This difference becomes visible when comparing reference data and simulations in Figure 73 for the last four reference points.

parameter	fixed parameters	varied parameters			
		minimum	maximum	ln(min.)	ln (max.)
φ (°)	20				
ψ (°)	0				
c (kPa)	27				
λ^*		0.05	0.15	-3	-1.9
μ^*		0.0005	0.02	-7.6	-3.9
k (m/day)		1.0E-07	1.0E-03	-16.1	-6.9

Table 17: Search intervals for the isotropic compression test.

163

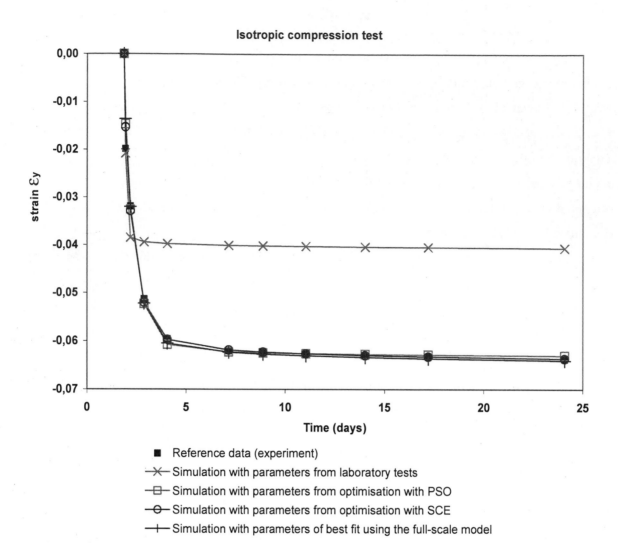

Fig. 73: Calculation results for the isotropic compression test.

(Laboratory test and simulation: this study; optimisation: *Meier 2009 U*).

Parameter	Experimental values	Identified values	
		PSO	SCE
φ (°)	20		
ψ (°)	0		
c (kPa)	27		
λ^*	0,057	0,080	0,085
κ^*	0,028	$(\lambda^*) / 2$	$(\lambda^*) / 2$
μ^*	1,2E-03	1,2E-03	3,0E-03
k (m/s)	2E-10	2,7E-11	3,1E-11

Table 18: Summary of parameter sets for the isotropic compression test.

(Laboratory test and simulation: this study; optimisation: *Meier 2009 U*).

Both identified parameter sets are presented in Table 18. The identified values of the coefficient of permeability k are one order of magnitude lower than the value determined in the laboratory. This might be attributed to the higher stress level – the permeability test was carried out at an effective stress of 200 kPa – causing a higher compaction degree of the soil.

4.7.1.4. Deviatoric creep test

After the isotropic compression test, the sample was left in the triaxial apparatus and a further test was carried out to investigate the creep behaviour of the material at a deviatoric stress state. Deviatoric stress states are assumed to be present in many places along the slope of Corvara: Normally, the soil bodies located along the shear-zones are subject to a nearly constant vertical load resulting from their overburden. By the movement of landslide masses situated downslope of them, e.g. by their detachment at detachment zones, these soil bodies are unloaded laterally, towards a state of active earth pressure. If in contrast, a part of the earth-slide is pushed forward by an upslope landslide mass that moves faster, lateral load increases towards a state of passive earth pressure. The above-described unloading causes creep-deformations and the more a soil body is unloaded laterally, the higher is the resulting deviatoric creep rate.

In the deviatoric creep test, the sample was unloaded laterally, from the isotropic stress state at 800 kPa, down to 650 kPa, while the vertical load remained constant at 800 kPa. Unloading was carried out within approximately one hour (53 minutes) and by steps of 10 kPa. After this relatively short unloading phase the sample was left creeping for 12 days. The displacements of the sample top were recorded continuously. Reference data points were taken shortly before the beginning of the unloading phase and then 1.5 hours, 1 day, 4.5 days, 8, 12 days after the end of the stepwise unloading.

To save calculation time, a separate numerical model was built for this laboratory test instead of adding calculation phases to the model of the isotropic compression test. In the model of the deviatoric creep test, the isotropic compression at 800 kPa was summarised to one calculation phase, because there was no need for simulating the deformations at several points in time within this phase of the test. Only the deformations at the end of the isotropic compression have to be reproduced correctly. The numerical model was built analogue to that of the isotropic compression test. The full-scale version was discretised with 1464 triangular 15-node elements, whereas for the model to be used in the optimisation routine, a configuration with 58 elements of the same type proved to be sufficient. This relatively low number arises from the fact that, contrary to the model of the isotropic compression test, for the simulation of the deviatoric creep test, the time-dependent course of primary consolidation (taking place during the first hours and days of isotropic compression) is not required to be modelled with the maximum possible accuracy. Loads were defined as they were applied in the laboratory but instead of the stepwise reduction of the cell pressure in the experiment, the lateral load was reduced in four calculation phases (from 800 to 770, 730, 690 and finally 650 kPa). Within these phases, the load level was decreased continuously in such a manner that the integral of the load as a function of time equals the test conditions.

Like for the isotropic compression test, a first-trial forward calculation was performed, using the average values of the parameters from laboratory tests presented in Table 14 (page 156) and Table 16 (page 160). The result of this calculation is presented in Figure 75 (page 168) in comparison to the reference data set. Like the reference data, after the unloading, the simulation shows a nearly linear development of the deformations with time but the amount of these creep deformations is, by far, underestimated.

Statistical analysis

A Monte Carlo procedure was performed on the same configuration of parameters as in Section 4.7.1.3. (page 161) for the isotropic compression test, i.e. on λ^* (or κ^*, for κ^* was set to 0.5 λ^*), μ^* and k. The scatter plot matrix is presented in Figure 74. Parameters were varied within the search intervals given in Table 19.

parameter	fixed parameters	varied parameters			
		minimum	maximum	ln(min.)	ln (max.)
φ (°)	20				
ψ (°)	0				
c (kPa)	27				
λ^*		0.01	1	-4.6	0
μ^*		0.0005	0.05	-7.6	-3
k (m/day)		1.0E-07	1.0E-03	-16.1	-6.9

Table 19: Search intervals for the deviatoric creep test used for the statistical analysis and for the optimisation procedure.

The matrix in Figure 74 is based on 1000 calls, showing all 366 parameter combinations with objective function values smaller than $2*10^{-10}$. The parameter λ^*, which here stands also for κ^* (in the fixed ratio of $\kappa^* = 0.5 \lambda^*$), and therefore describes the stress-dependent stiffness during loading and unloading, as well as the modified creep index μ^* show clearly marked minimum regions in their objective function projections. The objective function projection of the permeability parameter k presents no clear peak. This is because the simulation results are sensitive to k only at the beginning of the laboratory test, when as a consequence of the unloading, negative excess pore pressures are building up.

Therefore the statistical analysis shows that good deviation values can be obtained for quite different values of k; except for values from the left half of the interval plotted in the matrix of Figure 74. This

166

is because too low values of k lead to an overestimation of the amount of negative excess pore pressures and of the time they need for dissipation. Nevertheless, it was decided to vary also k during the optimisation instead of fixing its value. The latter would mean to enter a very rough estimate of k which might be wrong and therefore would not improve the quality of the calibrated model. In this type of relatively inhomogeneous materials, permeability can show large variations due to local differences in sample composition and therefore the value of k could be affected even by very small local changes in pore geometries that might occur due to the expansion of the sample during unloading.

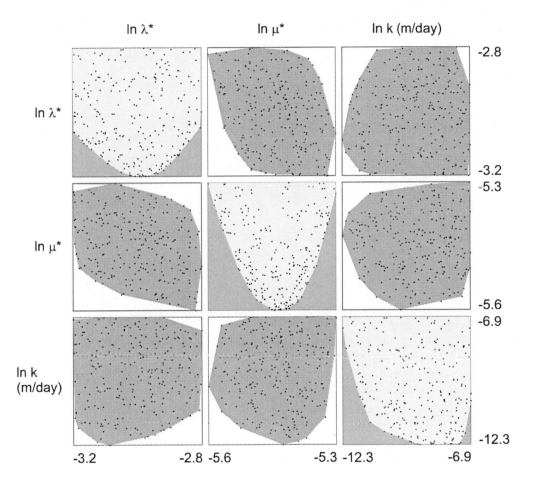

Fig. 74: Scatter plot matrix for deviatoric creep test (modified after *Meier 2009 U*).

Optimisation using PSO and SCE

After 500 calls, the Particle Swarm Optimiser had found a parameter set with a deviation value of $7.4*10^{-11}$, which could not be improved over several cycles. The SCE algorithm reached a similar result (deviation $7.3*10^{-11}$ with 377 calls of the forward calculation. The corresponding simulation results, which are nearly identical, are illustrated in Figure 75.

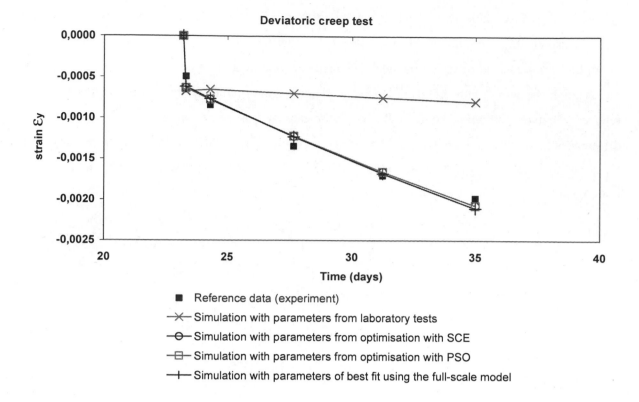

Fig. 75: Calculation results for the deviatoric creep test.

(Laboratory test and simulation: this study; optimisation: *Meier 2009 U*).

The identified parameter sets are given in Table 20. The values found by the different algorithms are nearly identical for λ^* (=2 κ^*) and μ^*, but for the reasons explained above, they differ for k. Furthermore, the identified values of μ^* suggest that much higher values of this parameter than the ones determined manually or the ones identified for the oedometer tests are needed to reproduce the amount of creep deformations under the stress conditions of the deviatoric creep test.

Parameter	Experimental values	Identified values	
		PSO	SCE
φ (°)	20		
ψ (°)	0		
c (kPa)	27		
λ^*	0,057	0,051	0,050
κ^*	0,028	$(\lambda^*) / 2$	$(\lambda^*) / 2$
μ^*	1,2E-03	4,4E-03	4,4E-03
k (m/s)	2E-10	3,8E-10	5,1E-09

Table 20: Summary of parameter values for the deviatoric creep test.

(Laboratory test and simulation: this study; optimisation: *Meier 2009 U*).

All in all, the simulations of the laboratory tests showed that the Soft Soil Creep model correctly simulates the material behaviour of the fine-grained soil-matrix of the landslide material and that the Inverse Parameter Identification Technique is able to find parameter sets that allow for an approximate reproduction of this behaviour under the boundary conditions and in the stress ranges of the experiments.

4.7.2. Discretisation of the numerical slope model

The discretisation of the numerical model of landslide source area S3 is illustrated in Figure 76. A plane-strain geometrical configuration with the real dimensions of the slope was used. Part A) of Figure 76 shows the Finite Element model used for the optimisation. The full-scale version of the model is shown in part B).

The full-scale model was discretised with 2790 triangular 15-node elements. Horizontal fixities were assigned to the lateral boundaries and total fixities to the basal boundary. The basal boundary was furthermore defined as impermeable. For the use in the optimisation procedure, the materials below the basal shear-zone, i.e. the older landslide material and the bedrock, which are assumed not to be involved in the landslide, were cut away and the lower boundary of the basal shear-zone was defined as fixed and impermeable. In addition, the model was cut above the upper end of the basal shear-zone, i.e. upslope from the 2D-projections of GPS-points 29 and 30, because this region, lying farther above the main detachment zone, was assumed not to be concerned by the movements of the main earth-slide of source area S3 (compare Section 4.3.2., page 128). The reduced model version was discretised with 2076 triangular 6-node elements. By means of the above described simplifications, the average calculation time could be reduced to less than a quarter of an hour on a standard personal computer, while the same computer needed several hours for one forward calculation with the full-scale model. Both Finite Element models were meshed with the automatic meshing procedure of the software, but along the shear-zones, mesh size and element shape were controlled manually by predefining geometry points. This was done to assure a sufficiently fine mesh with a suitable orientation of the triangular elements in order to avoid an excessive distortion of the element shapes by the expected calculated deformations. For a detailed illustration of the discretisation of the shear-zones, see part A) of Figure 76.

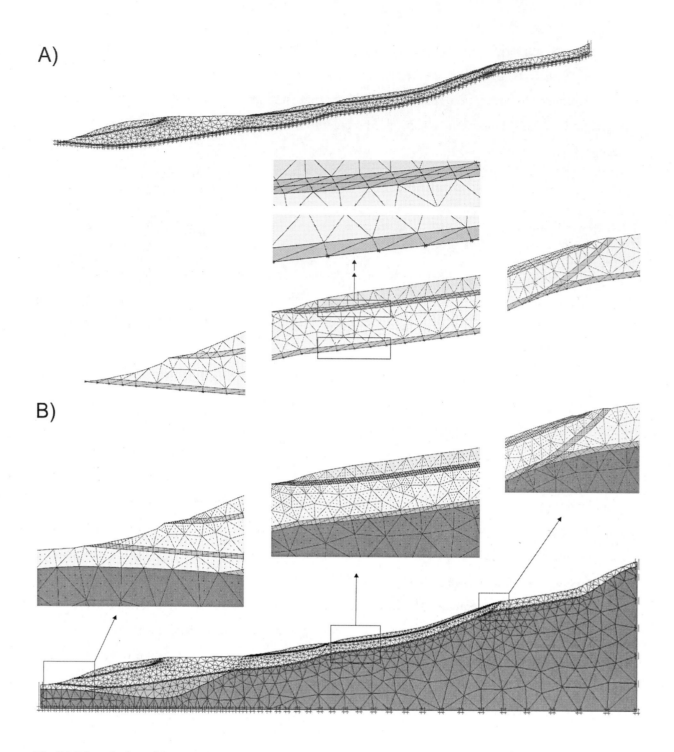

Fig. 76: Discretisation of the numerical slope model.

4.7.3. Calculation phases for generating the initial stress state of the simulation

The initial stress state at the beginning of the modelled time span, i.e. in the year 2001, is mainly a consequence of the quaternary geological and geomorphological history of the studied slope. Although detailed data on these topics was reported by different authors (see references in Section 4.2.2., page 119), the exact loading history of a slope is never known: The more one goes back in time, the coarser and the more uncertain is the information available with regard to the real geometries, pressures and material properties. Therefore, the initial stress state in the model was generated by simulating a very simplified "substitute" loading history, based on the available knowledge about the slope evolution. Modelling of the different phases of this simplified slope evolution was performed by means of a Staged Construction procedure (compare Chapter 3, page 87). During the generation of the initial stress state, the updated mesh option was deactivated, i.e. first-order theory was used.

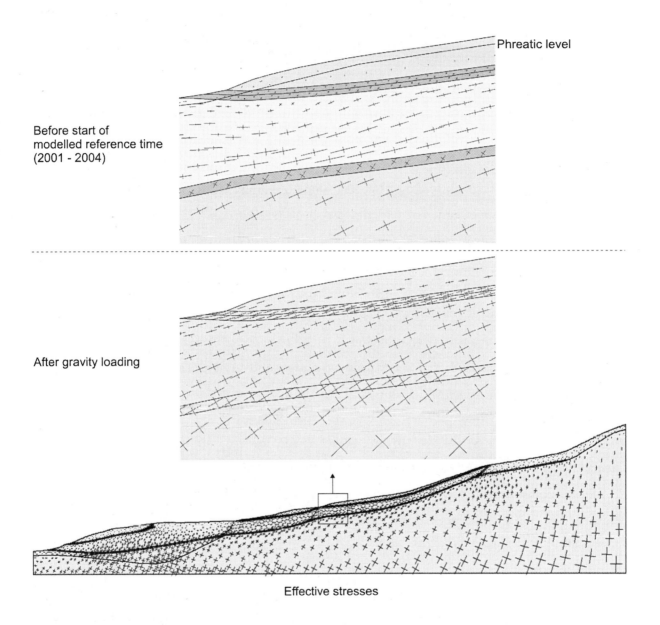

Fig. 77: Stress distributions during the generation of the initial stress state.

It is assumed that already before the last glaciation, the slope was characterised by a similar gently inclined profile. For this reason, the generation of the initial stresses should, from the beginning, take into account the influence of this topography on the distribution and direction of stresses (*Spickermann 2008*). Therefore, in a **first calculation phase**, a stress field was created by gravity loading based on the actual topography and assuming a homogeneous linear elastic material whose specific weight equals that attributed to the actual bedrock. During gravity loading, Poisson's ratio was adjusted to result in a ratio between horizontal and lateral stresses similar to that corresponding approximately to a K_0-value of 0.8. After the generation of the gravity loading stress field (see Figure 77), the value of Poisson's ratio was changed to the assumed real value (of 0.33), which was from then on used in all further calculation phases.

Landslide layers have formed mainly by the weathering of the bedrock formations (see Section 4.2.1., page 118). Thereby density and stiffness of the bedrock, whose diagenesis had taken place under much higher pressures than those acting in the first 200 m below the ground surface, have decreased (and continue decreasing) until ultimately the material transforms into a soil. At the same time, the most weathered parts of this material have become unstable (and continue becoming unstable) and were eroded. These processes have caused (and still are causing) a continuous unloading of the slope. Therefore, in a **second calculation phase** the slope model was unloaded by lowering the specific weight of all layers except for the bedrock to the assumed present-day values.

The stress field at the beginning of the Holocene slope instability is assumed to have formed during the unloading from the overburden of overlying ice-sheets (see Section 4.2.2., page 119). Therefore, in a **third phase**, a distributed load of 1000 kPa was applied perpendicularly to the ground surface of the model and in a **fourth phase**, the soft shear-zone layers were inserted into the model.

As during the Holocene the shear-zone materials were subject to disturbance by movements and alteration by weathering, they are assumed to have reached a near-normally consolidated state by now (see also Section 4.4., page 131, and Section 4.7.1.1., page 152). Consequently, in the model, the shear-zone layers were not allowed to compact and preconsolidate significantly under the overburden that was ascribed to the ice cover. Therefore, in a **fifth phase**, the slope was unloaded very fast to its actual state immediately after the shear-zone layers had been inserted.

The slope of source area S3 is known to have been affected down to a similar depth as today by earth-slide and earth-flow phenomena for the last 2500-3000 years (for details see again Section 4.2.2., page 119, and Section 4.4., page 131). During this time, the positions of the shear-zones were not constant. Therefore the loading history of the actual shear-zone domains is largely unknown and a reconstruction of this loading history in the model is practically impossible. For this reason, loading history of the shear-zones in the model was defined according to the hypotheses made in Section 4.4: (on page 131): In a **sixth phase**, after the unloading phase, the shear-zone materials were left creeping under the assumed average water pressures until approximately constant displacement rates were observed with respect to the time-scale of the simulated monitoring period. The duration of this phase (10000 days, approx. 27 years) had to be defined arbitrarily to comply with this criterion. The stress state reached at the end of this last phase of the alternative loading history was taken as initial stress state for the simulation of deformations in the monitoring period. Figure 77 illustrates the distribution of the effective stresses and their principal directions after the first and after the last of the previously described phases.

4.7.4. Calculation phases for the simulation of the monitoring period 2001-2004

The monitoring period between 2001 and 2004 was divided into calculation phases of rising, high, falling and low pore water pressures according to Section 4.5.6. (page 139). In order to calculate displacements at the dates of the measurements, these phases had to be split. To reduce the number of calculation phases and consequently calculation time, the time series of the reference data was reduced by five points. These data points will be displayed together with the results but they were not used for model calibration.

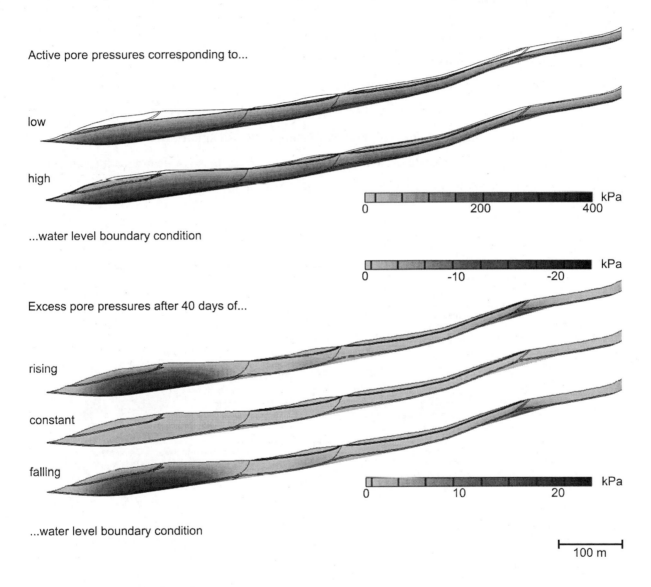

Fig. 78: Modelled distributions of pore water pressures.

Calculations were performed as consolidation analyses, this means taking into account the time-dependent development of pore pressures. Phases are modelled as staged construction phases defining a starting condition and a final condition for each of these phases. Within the phases of rising and falling water level, there is a continuous change of the pore pressure boundary condition, while within the phases of high and low water level, the respective boundary condition remains constant. Pore water pressures throughout the slope model develop depending on the pore pressure boundary conditions, the permeability of the materials and the deformations. Figure 78, in its upper part, shows the distribution of pore pressures as it was defined by the boundary conditions for the **low** and for the **high** water level scenario.

The lower part of Figure 78 illustrates the positive or negative excess pore pressures with respect to these boundary conditions as they are observed in the model at the end of a 40-day phase of falling or rising water level. As expected in Section 4.5.6. (pages 139 to 142), after the 40-day **rise**, the excess pore pressures are near zero for the most part of the shear-zone domains. It can be recognised, though, that pore water pressures in deeper regions, especially in Block C4, are still up to 20 kPa lower than those corresponding to the high pore pressure scenario. Vice versa, after the 40-day **fall**, positive excess pore pressures are still remaining. Thus, analogue to the real slope, a change of the pore pressure boundary condition has a delayed effect on the shear-zones. The larger the thickness of the overlying landslide- layers, and the smaller their permeability, the longer is this delay. The intermediate graphic in the lower part of Figure 78 shows the situation after another 40 days of **constant** water level boundary condition. Now the excess pore pressures are near zero throughout the entire landslide section and the low or high pore pressure scenario has fully taken effect.

4.7.5. Assumed material parameters of landslide-blocks and -layers and of the detachment zones

	Stress range due to overburden	Oedometer modulus of fine-grained matrix E_{oed}	Assessed value E_{oed} for entire soil mass	Fixed model parameter E_{ref}
	(kPa)	(MPa)	(MPa)	(MPa)
Shallower landslide layers	100 - 200	5 - 30 (4)	30	20,2
Main landslide body	400 - 800	11 - 53 (4)	60	40,5

Value ranges and number of samples (in parentheses)

	Geological Strength Index GSI (*)	Flysch type (*)	Sigma ci arenites (*)	Sigma ci pelites (*)	Modulus Ratio (*)	Calculated Rock Mass Modulus E_{rm}	Fixed model parameter E_{ref}
			(MPa)	(MPa)		(MPa)	(MPa)
Bedrock formations							
Outcrop data	30 - 45 (**)	B - D (**)	70 - 130 (**)	20 - 25 (**)			
Average conditions	35	C	55		200	1250	
Worst possible conditions	20	thick pelite-dominated zone		20	200	185	185

* Classified after *Marinos & Hoek 2001* and *Hoek & Brown 1997*
** From *Panizza et al. 2006*, 5 outcrops in La Valle Formation, 1 San Cassiano Formation,

	Concluded from appearance in borehole cores:	Estimated E_{ref}
Older landslide layer (no data)	E_{ref} (main landslide body) < E_{ref} (older landslide layers) < E_{ref} ("worst" rock masses)	100 MPa

Table 21: Derivation of stiffness parameters from laboratory data (this study) and outcrop data (*Panizza et al. 2006*).

Table 21 summarises the derivation of the assumed values of the stiffness parameters for the shallower landslide layers, the main landslide body, the older landslide layer and the bedrock from laboratory tests and outcrop data.

Shallower Landslide Layers and Main Landslide Body

The linear-elastic reference stiffness of the shallower landslide layers and of the main landslide body was defined based on the results of oedometer tests performed on four different samples taken from the fine-grained matrix (fraction < 2 mm) of the landslide material. Unloading and reloading were assumed to be the predominant loading cases, both during the detachment of landslide blocks from a more stable upslope region and during loading and unloading of the grain skeleton by the recurring fluctuations of the water table. Stiffness in the deeper parts of these layers, i.e. directly above the shear-zones was considered to be most relevant for the definition of the constant reference stiffness values. Therefore the assessment of the relevant stress-range was based on the pressure generated by an overburden of 5-10 metres for the shallower landslide layers and of 20-40 metres for the main landslide body. Stiffness parameters were then selected near the upper margin of the value ranges found in the oedometer tests, because, on a larger scale, the undeterminable quantity of coarser components, i.e. fragments of rock and rock masses, which are embedded in the matrix and, compared to this matrix, do not deform significantly, are assumed to augment the overall stiffness of these layers considerably.

Bedrock and Older Landslide Layers

The stiffness parameters of the bedrock and of the older landslide layers below the basal shear-zone are of little relevance for the simulation results. This was proven by the results of the reduced model, where the lower boundary of the basal shear-zone was just fixed without modelling these layers. However, in order to verify that actually these layers are not affected by the deformations, they were modelled in the full-scale version. Stiffness of the bedrock formations was estimated based on the classification data of six outcrops – which was published by *Panizza et al. 2006* – using the methods proposed by *Hoek and Diederichs 2006, Marinos and Hoek 2001* and *Hoek and Brown 1997*. The values for the uniaxial compressive strength of the intact rock pieces (Sigma ci) of the arenite layers in Table 21 appear to be quite high, when compared to laboratory values of uniaxial compressive strength reported in literature for comparable rock-types (see also the remarks in Chapter 3, page 60). Anyway, as it is shown also in Table 21, these values did not go into the derived stiffness parameters, because, considering the pronounced interbedding of very different strata in the La Valle and San

Cassiano formations, a stiffness value determined for their weakest parts was used in the model. Based on its appearance in the borehole cores, stiffness of the older landslide layer was assumed to be in the medium range between that of the bedrock in worst possible conditions and that of the main landslide body.

Detachment Zones

In the field, the detachment zones represent sets of neighbouring tension cracks. The shape of these cracks is listric and with depth they transform into shear surfaces which finally encounter in the related shear-zone. In the continuum mechanical model, detachment zones are simulated as zones, two metres thick, whose shape is similar to that of one of the above-described detachment surfaces and which consist of a material similar to the shear-zone materials. Like the shear-zone materials, this material can not resist any tensile stresses (i.e. tension cut off is set to zero) and its cohesion is negligible. Due to their upward transition into tension cracks with rather rough surfaces, the average friction angle of the detachment zones is assumed to be significantly higher than the friction angles of the shear-zones. Based on the above considerations, a friction angle of 27° and a cohesion value of five kPa were assigned to all detachment-zones and the rest of their material parameters were set equal to those of the corresponding shear-zones.

4.7.6. First-trial forward calculation using experimental parameter values for the shear-zones

The appropriate assessment of the material parameters of the shear-zones is very difficult, because the real stress history and exact present stress distribution are unknown and the complex geometry of the slope is simplified by the model. In the field, on larger scales than in the laboratory, the materials contain many discontinuities (mainly tension cracks and shear cracks) as well as chaotically arranged rock fragments of different sizes. Furthermore, the long-term shearing processes surely have introduced anisotropy into the shear-zone materials e.g. by the alignment of platy clay particles. The degree of this anisotropy is variable as a consequence of the heterogeneity of the materials and a representative estimate of this factor is practically impossible. All these inhomogeneities influence the overall material behaviour of the shear-zone layers, which in the model were simplified to continuous zones, each of them made up of one homogeneous and isotropic material.

In spite of these shortcomings, since the fine-grained soil matrix is assumed to dominate material behaviour of the shear-zones, a simulation with parameters obtained experimentally on the sample scale for this characteristic material should give a roughly and qualitatively correct system response with regard to the deformations of the slope. For this reason, a first-trial forward calculation using parameter values from the laboratory tests described in Section 4.7.1. (pages 148 to 169) was carried out. These parameter values were used for all shear-zones and are shown in Table 22.

material	parameter	unit	value
shear zones	*	none	0,057
	*	none	0,028
	μ*	none	1,2E-03
	°		20
	c (s3)	kPa	27

Table 22: Experimental parameter values used in the first trial forward calculation.

Cohesion was assumed to be near zero (0.3 kPa) as a result of the long-term displacements (see Section 4.4., page 131) except in shear-zone domain s3. For the location of shear-zone domain s3, see Figure 80 (page 179). In the field, in this area, the shallower landslide layers are at present moving mainly oblique to the direction of the 2D model section and are flanked by zones where this shallower landslide activity has calmed down. Therefore, the average cohesion value determined in shear tests on the landslide material (compare page 151) was used for shear-zone domain s3.

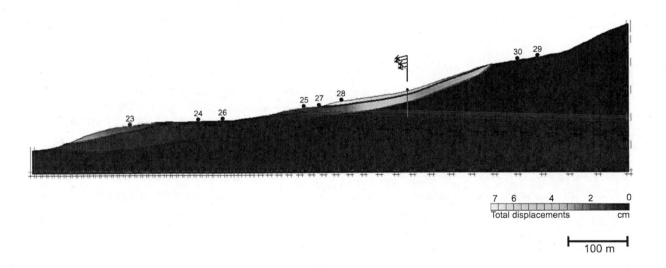

Fig. 79: Results of first trial forward calculation with the slope model.

Figure 79 shows the calculated total displacements at the end of the three years of the monitoring period. Deformations are, by far, underestimated, with a maximum of total displacements of around seven centimetres, and the region between GPS points 24 and 25, which is active in the field, did not move in the model. However, the relative amounts of displacements are qualitatively correct in many respects: Same as in the measurements, the largest displacements are observed at point 28 and they are much larger than those obtained for point 27. Displacements at point 23 are higher than at points 24 and 26, but lower than at point 28. Negligible displacements (smaller than 2 mm) were calculated for points 29 and 30, as well as for all regions below the basal shear zone. In the most active region, also the shape of the displacement profile and the direction of displacement vectors are in qualitative agreement with the measurements.

4.7.7. General settings of the statistical tests and the optimisation procedure applied to the slope model

4.7.7.1. Fixed parameters and varied parameters

The displacement rates along the shear-zones depend on shear strength and creep behaviour of these zones. Modelled creep behaviour is controlled mainly by the material parameter μ^* (modified creep index). As mentioned already in Section 4.7.1.3. (page 162), creep behaviour of natural soils is also related to their stress-dependent compressibility and vice versa. The latter is described by the model parameter λ^* (modified compression index). This means that λ^* and μ^* depend on each other. As both parameters and also their ratio under the conditions prevailing along the shear-zones of landslide source area S3 are unknown, it was decided to identify both μ^* and λ^* of these zones, varying them within large search intervals (Fixing λ^* at a wrong value could possibly impede a good calibration result).

Based on the considerations made in Section 4.7.6. (page 177), the cohesion, c, was set to a value near zero (0.3 kPa) for all shear-zone domains, except s3 (see Figure 80 for the location of s3), where a search interval between 5 kPa and 50 kPa was defined. Representative values of the friction angle φ of the shear-zones are assumed to be controlled by the residual strength of the landslide material's soil matrix and by the amount of roughness and obstacles caused by coarser components such as rock blocks. Consequently, the friction angles of the shear-zones were selected to be identified, varying them between 15° and 25°, i.e. the estimated residual friction angle of the soil matrix and the friction angle attributed to the bedrock formations when they are in a heavily weathered and disintegrated state.

The value of κ^* (modified swelling index) was fixed for all shear-zones based on laboratory data: values of κ^* obtained in five oedometer tests on different samples from the soil matrix of the landslide material (fraction < 2 mm) ranged between 0.01 and 0.05, with an average value of 0.026. Due to the influence of less deformable coarser components, representative values of κ^* for the shear-zones were assumed to be somewhat lower, i.e. stiffness is higher than that of the fine-grained fraction. Therefore, the value of 0.02 was used in the calculations.

4.7.7.2. Parameter links and parameter constraints

The results of the first-trial calculation suggest that the material parameters of the shear-zones are not equal for all shear-zone domains. Yet the basic engineering-geological hypothesis of this work, which was set up on the basis of the field investigations, postulates that on principle, all shear-zone materials are similar to each other, because of their very similar stratigraphical provenance and analogue quaternary geological genesis. Hence, some of the parameters of the shear-zones were linked to each other, while others were chosen to be identified independently, trying to minimise the number of variable parameters but not to oversimplify the model.

Parameter links

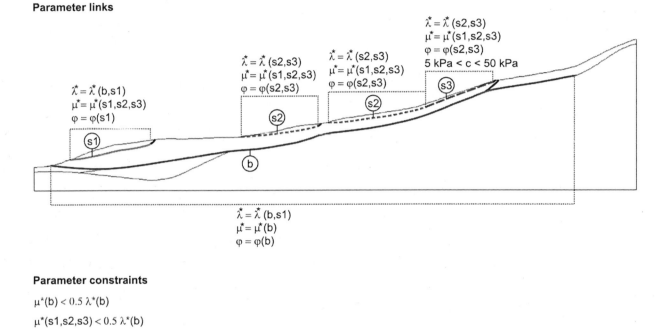

Parameter constraints

$\mu^*(b) < 0.5\,\lambda^*(b)$

$\mu^*(s1,s2,s3) < 0.5\,\lambda^*(b)$

$\lambda^*(s2,s3) > \lambda^*(b,s1)$

(b)   Material of basal shear zone

(s1)(s2)(s3)   Materials of secondary shear zones

⊢———————⊣
100 m

Fig. 80: Parameter links and parameter constraints defined for the statistical analysis and the optimisation procedure.

The shear-zone material of the two shallower landslide layers situated in the upper part of the slope (s2 and s3 in Figure 80) was assumed to be rather softer, certainly not harder than the material of the basal shear-zone. In the accumulation region near the foot of the slope, the material of the secondary shear-zone (s1 in Figure 80), besides being less compacted due to the smaller overburden, was regarded as very similar to that of the basal shear-zone. Therefore the modified compression index λ^* of s1 was set equal to that of the basal shear-zone (b), and, respectively, that of s2 equal to that of s3. Additionally, a parameter constraint, demanding for λ^* of the secondary shear zones s2 and s3 to be larger than that of

179

the basal shear-zone in all forward calculations, was set in order to save calculation time by excluding unrealistic parameter combinations that would overestimate the stiffness of the shallower shear-zones. As the value of the modified creep index $\mu*$ depends strongly on the stress level, $\mu*$ of the deep-seated basal shear-zone was expected to be different from the modified creep indices of the secondary shear-zones (s1, s2, and s3), which were assumed to be similar to each other. Consequently $\mu*$ of the basal shear-zone was selected to be identified separately from the modified creep indices of the secondary shear-zones, and $\mu*$-values of s1, s2 and s3 were set to be equal among each other. In order to keep the number of failed forward calculations low, two parameter constraints were prescribed to avoid physically and numerically unfavourable parameter combinations, demanding for the highest $\mu*$-value to be smaller than half of the lowest $\lambda*$-value in all parameter sets. A summary of all above-described parameter links and parameter constraints is given by Figure 80.

4.7.7.3. Weighted reference data set for statistical analyses and optimisation procedure

For the application of the inverse modelling method, the reference data were given weighting factors according to their precision, as described in Section 4.6.1. (page 143). For details on the measurements, see Section 4.2.4. (page 122). The horizontal displacements of GPS-points 23-28 were assigned a weighting factor of 1.0; the vertical displacements of these points were given a weighting factor of 0.2. Instead of the numerical values (which were in the range of the measurement error) zero displacements were used as reference data for GPS points 29 and 30. Their horizontal reference displacement was weighted by a factor of 0.5 and vertical displacement by 0.1. In order to assure high objective function values in the case of forward calculations that might show good fits for the measurement points at the ground surface but could be characterised by unrealistic displacement profiles along depth, two additional reference points were set in the model. These virtual measurement points were placed shortly above the basal shear-zone, exactly below GPS point 24 (i.e. near inclinometer C4) and below GPS point 27, which is assumed not to be affected by the shallower landslide phenomena (see Figure 57, page 129). Based on the interpretation of the inclinometer data explained in Section 4.3.3. (pages 129 and 130), displacements of these reference points at depth were defined as identical to those measured at the ground surface.

4.7.8. Statistical analysis on the parameters of the shear-zones

In spite of the linking of similar parameters, still eight parameters remained to be identified. A statistical analysis, based on 7670 calls of the forward calculation was carried out on these parameters, varying them within the intervals specified in Table 23 (page 182). The scatter-plot-matrix is presented in Figure 81. This matrix shows the parameter combinations related to the 83 best objective function values. None of the plots has a clear-cut shape, therefore the problem must be considered as underdetermined and it is advisable not to increase the number of varied parameters. However, the objective function plots give some information about the most probable ranges of the best parameter values.

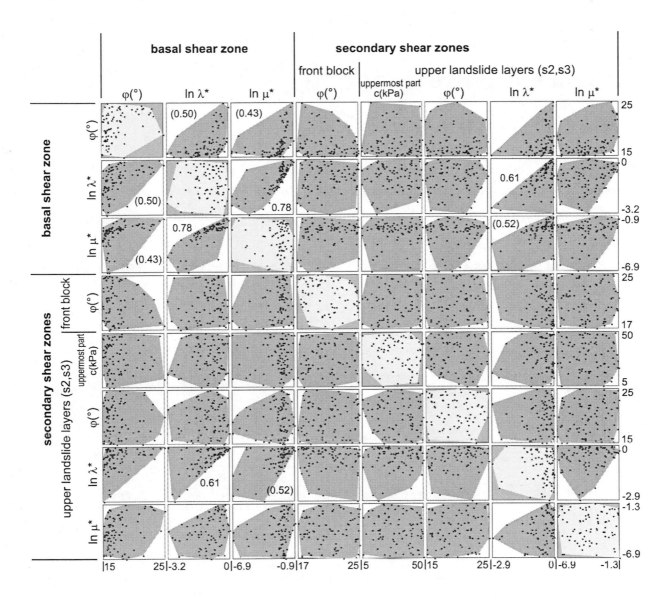

Fig. 81: Scatter plot matrix for the slope model (modified after *Meier 2009 U*).

For the friction angle φ of the **basal shear-zone**, the best parameter sets are all located in the left part of the search interval, i.e. between 15° and 19°, whereas for μ* of this zone, they accumulate in the right part of the interval. Less clearly, this can be noticed also for the values of λ*, because μ* and λ* appear to be correlated (correlation coefficient of 0.78), as previously expected. None of the 83 best parameter combinations has a value of μ* in the uppermost part of the search interval, i.e. between logarithmised μ*-values of -0.9 and -0.69.

All in all, the information obtained for the parameters of the **secondary shear-zones** is less clear, because they do not influence the quality of the fit at all reference points like the parameters of the basal shear zone (but only at GPS-points 23, 25 and 28; see again Fig. 57, page 129, for the Geometry Model). As in none of the displayed best parameter sets φ of the front block (s1) is below 17° and furthermore, in 15 of the 18 best sets, φ-values are between 18° and 23°, the statistical analysis indicates that an optimum value of this parameter can be identified in the medium range of the search interval. Good objective function values can be obtained independent of the cohesion value, c, assigned to shear-zone domain s3 in the uppermost part of the shallower landslide layers, except for values from the lowermost part of the interval. In contrast, relatively good fits are observed for a wide range of friction angles, φ, of this zone, but not for values from the uppermost part of the search interval.

In the case of λ* of the upper landslide layers, the accumulation of plotted points on the right side of the interval as well as the weak correlation between λ*(b) and λ*(s2, s3) are direct consequences of the parameter constraint demanding for λ*(s2, s3) > λ*(b). The objective function projection for the parameter μ* of these layers shows good model fits independent of the parameter value, except for the uppermost part of the search interval (between logarithmised μ*-values of -1.3 and -0.69) where none of the 83 best parameter sets plotted.

material	parameter	unit	fixed values	vary between				constraint
				min.	max.	ln (min.)	ln (max.)	
b basal shear zone	λ* (b)	none		0,04	1	-3,22	0	
	κ* (b)	none	0,02					
	μ* (b)	none		0,001	0,5	-6,91	-0,69	μ*(b) < 0,5 λ*(b)
	φ (b)	°		15	25			
	c (b)	kPa	0,3					
s1 secondary shear zone of front block	λ* (s1)	none		λ*(s1) = λ*(b)				
	κ* (s1)	none	0,02					
	μ* (s1)	none		μ*(s1) = μ*(s2)				
	φ (s1)	°		15	25			
	c (s1)	kPa	0,3					
s2 secondary shear zones of upper landslide layers	λ* (s2)	none		0,04	1	-3,22	0	λ*(s2) > λ*(b)
	κ* (s2)	none	0,02					
	μ* (s2)	none		0,001	0,5	-6,91	-0,69	μ*(s2) < 0,5 λ*(b)
	φ (s2)	°		15	25			
	c (s2)	kPa	0,3					
s3 secondary shear zone uppermost part	λ* (s3)	none		λ*(s3) = λ*(s2)				
	κ* (s3)	none	0,02					
	μ* (s3)	none		μ*(s3) = μ*(s2)				
	φ (s3)	°		φ (s3) = φ (s2)				
	c (s3)	kPa		5	50			

Table 23: Search intervals and fixed parameters used for the statistical analysis and the optimisation procedure.

From the scatter plot matrix in Figure 81 it is not clear whether the interval for λ* of the shear-zones of the shallower landslide layers (s2 and s3) is sufficient since the data points with low deviation values accumulated near the upper boundary. As for higher values of λ* (e.g. between 1 and 2,

182

logarithmised values between 0 and 0.69) the percentage of failed forward calculations increased considerably, it was decided not to enlarge the interval unless there would be clear evidence that best objective function values can be found only for the highest λ*-values located at the upper margin.

secondary shear zones

upper landslide layers (s2,s3)

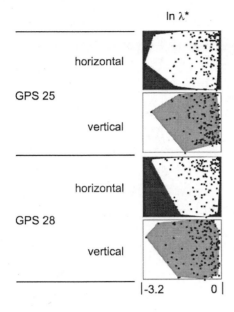

Fig. 82: Objective function projections for the displacements of GPS points 25 and 28 (data from *Meier 2009 U*).

Only the displacements of GPS-points 25 and 28 are directly influenced by the setting of this parameter, therefore, in a second analysis, objective function values were determined separately for the measurement series of these points. In Figure 82, the best 150 values of each of these four different objective functions are plotted against the parameter λ*. A closer look at these data shows that in none of the plots the best values were located at the boundary of the interval. For this reason, the unchanged search intervals of Table 23 were used for the optimisation procedure.

4.7.9. Identification of shear-zone parameters with different algorithms

First step of the optimisation procedure: complete reference data set

After 1010 calls of the forward calculation, the PSO algorithm was not able to further reduce the deviation from the reference data. With a similar number of calls (1100), the SCE algorithm achieved a significantly lower deviation ($9.2 * 10^{-4}$ against $2.8 * 10^{-3}$ reached by the PSO algorithm). During a high number of calls, this result could not be improved any more. The parameter sets identified by both algorithms, using the complete reference data set as described in Section 4.7.7.3. (page 180), are shown in Table 24.

parameter	unit	complete reference dataset		narrowed reference dataset	simplified model
		(PSO)	SCE	SCE	PSO
λ^* (b,s1)	none	(0.78)	0.62	fixed	0.93
λ^* (s2,s3)	none	(0.79)	0.96	0.63	
μ^* (b)	none	(0.39)	0.17	fixed	0.20
μ^* (s1,s2,s3)	none	(0.33)	0.31	0.24	0.35
φ (b)	°	(24.1)	18.1	fixed	18.9
φ (s1)	°	(21.4)	21.6	19.5	
φ (s2)	°	(24.1)	16.2	16.4	21.7
φ (s3)	°			23.8	
c (s3)	kPa	(48.7)	48.3	16.9	fixed

Table 24: Summary of identified parameter values for the slope models.
(Slope models and simulations: this study; optimisations: *Meier 2009 U*).

In Figure 84 (page 186), the results of a calculation using the best parameter set of this first step of the optimisation procedure can be compared to the measured horizontal displacements of GPS points 23-28. Considering the simplifications made in the model it can be stated that displacements are reproduced quite well for most points, but substantial underestimation of displacements occurred at reference points 23 and 28.

Second step of the optimisation procedure: narrowed reference data set

As these points both lie above secondary shear-zones (see Figure 57 on page 129), in a second step of the optimisation procedure, the parameters of the basal shear-zone were fixed at the identified best values. For this optimisation, a narrowed reference dataset was used. This dataset contained only the horizontal displacements of the three measurement points situated above secondary shear zones (reference points 23, 25 and 28). Horizontal displacements were used as they represent the data of highest quality.

For the search, the configuration of parameter links, parameter constraints and search intervals (of Table 23, page 182 and Figure 80, page 179) was modified in three details:
As both identified cohesion values for s3 (48.7 kPa and 48.3 kPa) lied near the upper boundary of the search interval, this interval was enlarged until 100 kPa. Secondly, the optimisation algorithm was allowed to vary the friction angle of this shear-zone domain independently of that of s2. This additional scope for the parameters of s3 was provided because a bad calibration in this part could

hamper also the identification of proper parameter values for the region of s2, where reference point 28 is located. Thirdly, when fixing the parameters of the basal shear-zone, the parameter constraint demanding for $\lambda^*(b) < \lambda^*(s2, s3)$ was removed because it was preferred not to restrict the search based on a value of an identified parameter set. This parameter set is considered to be just one of a number of possible approximate solutions, which was found to represent the reference data best, i.e. with the lowest deviation. Yet it does not contain "only true" values of each of the parameters. Thus, a total of six parameters were varied at the second step of the optimisation procedure. After 2092 calls, the optimisation was stopped because again no significant reduction of the deviation was observed during a large number of cycles. The identified best parameter set is given also in Table 24. The value identified for λ^* of shear-zone domains s2 and s3 (0.63) is nearly identical to the one found for the basal shear-zone and s1 (0.62), suggesting that, on average, the materials of all shear-zones might be very similar with respect to their compressibility and creep behaviour (both μ^*-values are around 0.2). Figure 83 presents the displacement vectors calculated for the three years of the monitoring period (2001-2004) in comparison with those determined by means of GPS measurements. Reproduction of the horizontal displacements is very good. Except for GPS points 26 and 28, vertical displacements are clearly overestimated, even if the low precision of the vertical displacement measurements is taken into account. However, the rough direction of the simulated displacements is sub- parallel to the slope as it is generally observed in all sectors of the Corvara landslide. Qualitatively, the modelled displacement profile at point 24 is in accordance with that measured by the nearby inclinometer C4 (see Figure 58 on page 130).

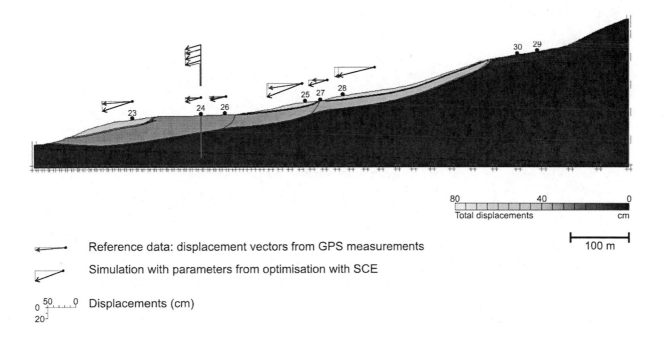

Fig. 83: Calibrated slope model – comparison of simulated displacement vectors with the reference data derived from the GPS measurements.

(Slope model and simulation: this study; GPS-data: *Corsini 2009 U*; optimisation: *Meier 2009 U*).

185

In Figure 84, simulation results are compared to the measured time series of the horizontal displacements at GPS points 23-28 (Displacements are shown at the same scale for all points).

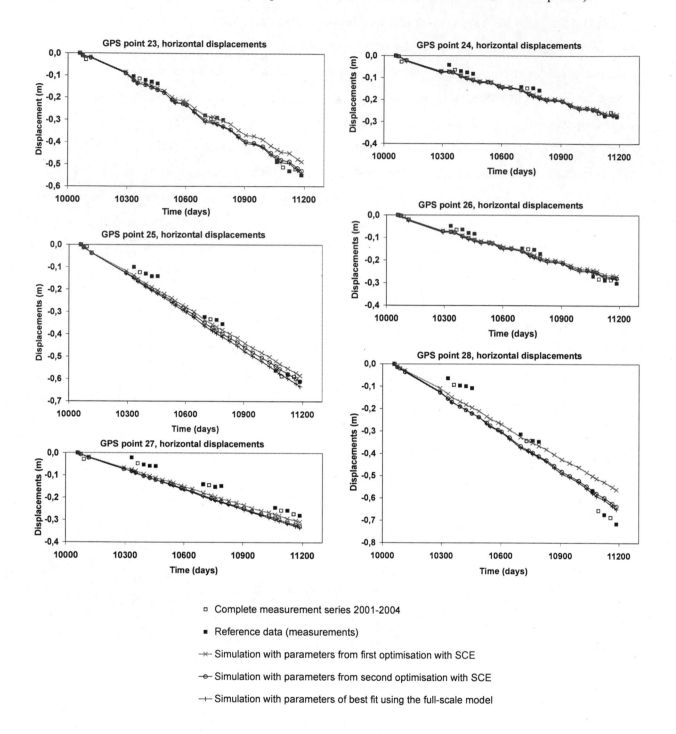

□ Complete measurement series 2001-2004

■ Reference data (measurements)

—×— Simulation with parameters from first optimisation with SCE

—○— Simulation with parameters from second optimisation with SCE

—+— Simulation with parameters of best fit using the full-scale model

Fig. 84: Time series of calculated displacements of GPS points 23-28 and measured reference data.
(Slope model and simulation: this study; GPS-data: *Corsini 2009 U*; optimisation: *Meier 2009 U*).

For points 23, 24 and 26, the simulated curves are close to the real time series, except for small differences, which are always below 5 cm. At points 25, 27 and 28, total horizontal displacements are reproduced approximately by the model, but displacements are overestimated in the phases of low water pressures and underestimated in the phases of high water pressures. This result is attributed to the fact that the modelling of the pore pressures had to be based on simplifying hydrogeological assumptions. In the case of the first group of points, these assumptions are apparently more appropriate for describing reality as in the case of the second group of points. In this context, it is probably not surprising that the measurement points belonging to the first group are located in that slope region where most of the hydrogeological data could be collected.

4.7.10. Simplification and adaption of the model for the pre-evaluation of countermeasures

All in all, the calibrated numerical model was regarded as sufficiently realistic to allow for the pre-evaluation of implemented countermeasures. For this type of practical applications, the model was adapted and simplified. For the pre-evaluation of mitigation measures, the values fixed for the stiffness parameters of the blocks and layers between the shear-zones were reduced. With regard to a dependable evaluation whether interventions can be effective, it makes little sense to take assumed average stiffness parameters as a basis, because a precise assessment of these values is very difficult. Earth-slide and earth-flow materials are very heterogeneous; it is also not known how the indeterminate number of discontinuities inside the landslide body influences its overall stiffness on larger scales. In any case, they represent potential additional paths for movements. In areas characterised by many discontinuities, such as shear cracks or tension cracks, material behaviour might be softer than expected. If the estimated stiffness of a landslide block is too high, the simulation will credit stabilisation measures which are successful locally with a higher range than they have in reality. Even if these rough estimates are correct, large uncertainties remain, because, due to the width of the affected slope area (of several 100 metres), the weaker portion of the heterogeneous materials will be decisive for the success of eventual stabilisation measures. Furthermore, when landslide bodies are moving towards newly built retaining walls or newly stabilised regions, (passive) earth pressures might locally exceed the previous maximum stress level. In this case, stiffness might be lower than the values estimated in Section 4.7.5. and Table 21 (page 174) from the unloading and reloading curves of the oedometer tests.

Due to the above considerations, the stiffness parameters for the shallower landslide-layers and the main landslide body were now chosen near the lower boundary of the intervals deduced from the laboratory tests. Oedometer Moduli were fixed at 5 MPa and 10 MPa, respectively. When compared to the stiffness parameters obtained during first loading, these values are in the medium range: Between the load steps of 100 kPa and 200 kPa, five oedometer tests on different samples from the matrix of the landslide material (fraction < 2 mm) gave Oedometer Moduli between 2 and 10 MPa, and for the load steps of 400 kPa and 800 kPa, computed moduli were between 5 and 18 MPa (in four different tests).

Instead of the detailed modelling of phases of low, rising, high and falling pore water pressures, an assumed representative pore pressure distribution was used as constant input for the simulation of the monitoring period. For this purpose, the pore pressure situation calculated in Section 4.5.4.2. (page 136) by means of a steady state groundwater flow calculation based on the assumed average water level of Section 4.5.2. and Figure 59 (page 134) was used. The model, which then produced constant displacement rates, was calibrated by means of the average displacement rates measured between 2001 and 2004. In this way, the simplified model is independent of the additional assumptions regarding the seasonal oscillations of the water level and – also by means of a groundwater flow calculation – the effect of countermeasures on the average hydrogeological situation can be evaluated more easily.

Furthermore, the material properties of all secondary shear-zones were set to be identical. Based on the general hypothesis of Section 4.7.7.2. (page 179), which was confirmed by the identified best parameter set (see, Section 4.7.9. and Table 24, pages 185 and 184), the modified compression index λ^* of the secondary shear-zones was set to be equal to that of the basal shear-zone (assuming similarity of these materials). Consequently, only five parameters were varied in the calibration process. Search intervals, links between the parameters and parameter constraints can be viewed in Table 25.

material	parameter	unit	fixed values	vary between				constraint
				min.	max.	ln (min.)	ln (max.)	
b basal shear zone	$\lambda^*(b)$	none		0,01	1	-4,61	0	
	$\kappa^*(b)$	none		$\kappa^*(b) = 0{,}5\ \lambda^*(b)$				
	$\mu^*(b)$	none		0,001	0,5	-6,91	-0,69	$\mu^*(b) < 0{,}5\ \lambda^*(b)$
	$\varphi=(b)$	°		15	25			
	$c=(b)$	kPa	0,3					
s secondary shear zones	$\lambda^*(s)$	none		$\lambda^*(s) = \lambda^*(b)$				
	$\kappa^*(s)$	none		$\kappa^*(s) = 0{,}5\ \lambda^*(b)$				
	$\mu^*(s)$	none		0,001	0,5	-6,91	-0,69	$\mu^*(s) < 0{,}5\ \lambda^*(b)$
	$\varphi=(s)$	°		15	25			
	$c=(s)$	kPa	0,3					

Table 25: Search intervals and fixed parameters used for the calibration of the simplified model of source area S3.

The model was then calibrated analogue to the procedures described in the previous Sections. For identifying material parameters of the shear-zones, the PSO algorithm was applied. After 1000 calls of the forward calculation, deviation had been reduced to a low value and could not be lowered significantly during several cycles. The identified best parameter set is presented together with the previous results in Table 24 (page 184). The displacement vectors calculated for the years 2001-2004 are illustrated together with the reference data in Figure 85.

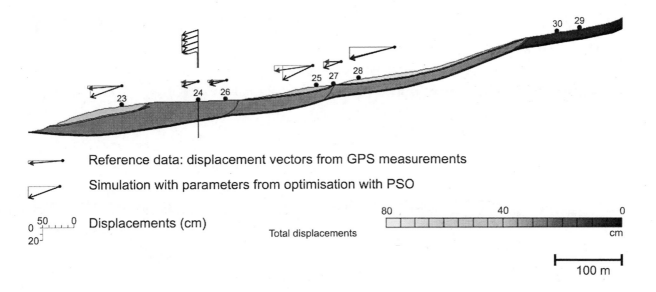

Reference data: displacement vectors from GPS measurements

Simulation with parameters from optimisation with PSO

Displacements (cm)

Total displacements

80   40   0

cm

100 m

Fig. 85: Results of the calibration of the simplified slope model.
(Slope model and simulation: this study; GPS-data: *Corsini 2009 U*; optimisation: *Meier 2009 U*).

4.7.11. Evaluation of the efficiency of implemented mitigation measures

The effectiveness of different combinations of countermeasures was then evaluated by implementing them into this calibrated simplified model (*Knabe et al. 2009*). Some of the most important results of the study of *Knabe et al. 2009*, which was based on the results of this thesis, are shortly presented in this Section.

Combination of slope-reinforcement and drainage measures

Figure 86 shows in its upper part (1-3) one of the investigated combinations of slope-reinforcement and drainage measures. The construction of three anchored retaining walls across the slope of source area S3 was simulated by inserting elastic plate elements (beam elements in the 2D-model) and elastic linear anchor elements into the model, within a staged construction phase of two weeks. Plate elements were modelled as fixed at the lower boundary of the basal shear-zone and also the lower ends of the anchors, whose inclination was 25°, were modelled as fixed.

Stiffness parameters of plate and anchor elements were calculated on the basis of the dimensions of structures built for the stabilisation of the Ca'Lita landslide (a large complex rock-slide/earth-slide/earth-flow) in the Northern Apennines, as reported by *Sartini et al. 2007*, *Corsini et al. 2006* and *Borgatti et al. 2008*. For the design to be realised in the field, non-contiguous bored pile walls, with a concrete wall serving as pile cap beam, were assumed. Distance between the piles was set at four metres and pile diameters at 1.4 metres. All pile heads were assumed to be tied back by anchors.

189

Finally, a normal stiffness of $7.5*10^7$ kN/m and a bending stiffness of $1.0*10^7$ kN/m² were assigned to the plate elements and an axial stiffness of $2.0 * 10^7$ kN/m to the anchor elements.

Investigated combination of slope-reinforcement and drainage measures

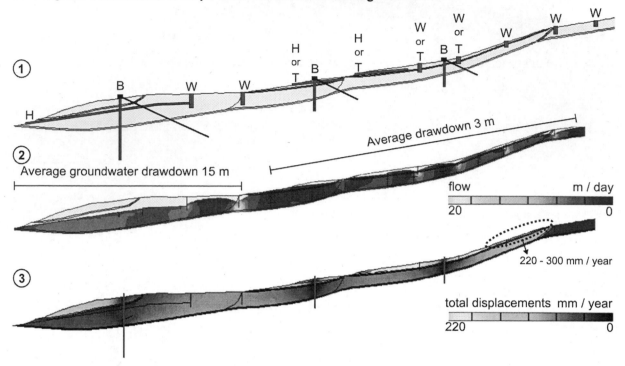

Number of similar interventions required to stop the displacements (approximately)

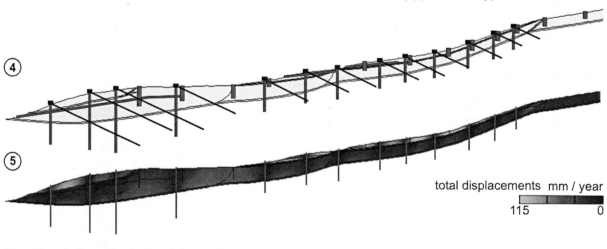

B   Bored pile walls, tied back by anchors

W   Drainage wells
H   Horizontal bored drains
T   Drainage trenches, deep, filled with permeable material

100 m

Fig. 86: Implementation and evaluation of mitigation measures (data from *Knabe et al. 2009*).

Contrary to the case of Ca'Lita, where bedrock could be encountered at 9-10 metres depth in some landslide sectors, in the case of Corvara, due to the very thick and quite soft layers of soils, pile lengths and anchor lengths of several tens of metres are needed to obtain the simulated fixed supports approximately in practice. In Figure 86, pile and anchor lengths as they are expected to be required at the different locations are drawn to scale, assuming that the structures are embedded ten metres deep into bedrock layers.

In combination with the retaining walls, a drawdown of the groundwater level by means of drainage measures was simulated. Drainage wells, horizontal bored drains and drainage trenches were modelled as lines, along which pore water pressures were set equal to zero during a steady state groundwater flow calculation that was carried out analogue to Section 4.5.4.2. (page 136). In the model, an average drawdown of 15 metres in the foot area and of 3 metres along the upper part of the slope, including the main detachment zone, could be achieved with the configuration of drainage measures illustrated in part 1 of Figure 86. The resulting flow-field and water level are shown in part 2 of Figure 86.

Total displacements calculated for the three years following the installation of the slope reinforcement and drainage measures, are displayed in part 3 of Figure 86. Considerable diminution of displacement rates is observed only in the neighbourhood of the retaining walls. Along the rest of the slope, average displacement rates still reach up to 20 centimetres per year. With respect to the current site-specific relevance of Source Area S3 (see Section 4.2.3., page 121), this result must be regarded as unsatisfactory. Furthermore, the achieved stabilisation effect seems be too small to protect the wells and drains from being destroyed by the movements. Part 4 and part 5 of Figure 86 show the number of interventions similar to those presented in part 1 that would be required to stop the displacements approximately in the model. Along the most relevant slope section – between foot-zone and main detachment-zone –, which is around 800 metres long, 13 bored pile walls and a similar number of series of wells had to be simulated. Obviously, due to the enormous construction effort and high costs, this configuration does not represent a realisable option.

Only drainage measures

Figure 87 shows the development of the simulated maximum displacement rates and the calculated overall factors of safety with time if the drainage measures of Figure 86, part 1 and 2, are implemented without any further interventions (i.e. no walls, no anchors). Factors of safety were calculated using the Phi-c-reduction method (*Reference Manual PLAXIS V8 2003*). Starting from a situation defined by the end of a calculation phase, the strength parameters, tan φ (tangent of the friction angle) and c (cohesion) of the soils are successively and proportionally reduced until driving forces exceed the holding forces. The Mohr-Coulomb failure criterion and constant reference stiffness values, determined on the basis of the stress-dependent material stiffnesses at the starting point, are used in these elastic-plastic calculations. The Phi-c-reduction factor of safety is then defined as the ratio between "available strength" and "strength at failure", i.e. as the ratio between the input value of tan φ and the reduced value of tan φ, and just as well between the input value of c and the reduced value of c.

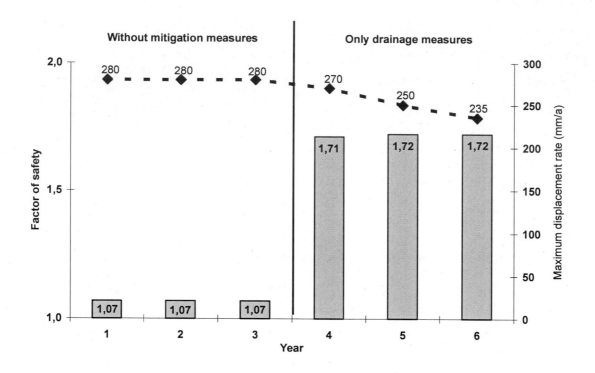

Fig. 87: Effect of successful lowering of water pressures on the simulation (data from *Knabe et al. 2009*).

During the first three years, i.e. in the simulated monitoring period, maximum total displacement rates are constant and the calculated factors of safety are close to one. This value is considered to give a realistic representation of the situation in the field, because the constant average movement rates during several years (2001-2008) are interpreted as the result of a balance between driving forces (due to gravity) and holding forces (due to friction and viscous resistance). As during these three years the geometry of the slope model did not change significantly and constant water conditions were simulated, factors of safety calculated for the end of each year remain (approximately) constant. After the third year, the groundwater level was lowered to the level depicted in Figure 86 part 2 (page 190). A prominent increase of the calculated factors of safety is then observed and maximum displacement rates start decreasing immediately (i.e. already in the fourth year of the simulation). After three years with water pressures kept at the lowered level, displacement rates are still high, but – for the simulation – the positive effect of the drainage measures is evident.

4.8. Discussion of results

The results of the simulations indicate that a short-term stabilisation of the slope of source area S3 by means of structural countermeasures might probably not be achieved. The combination of measures illustrated in part 1-3 of Figure 86, despite implying high financial effort, is unlikely to show a significant effect within a few years. When scheduling also some climatically adverse years (compared to the assumed normal state of the monitoring period), it can not be excluded that additional costs for repair works on the structures could arise before any positive effect would be noticeable. Considering the dimensions of the Corvara landslide and the related phenomena, any type of expensive measures should be discarded if there is no clear indication for their effectiveness: Besides source area S3, there are three other source areas (S1, S2, S4, see again Fig. 52, page 120), and for none of them it can be guaranteed that in the medium term (i.e. in the decades to come) no significant accelerations will occur. However, the simulations suggested that, in the medium term and on average, countermeasures, even of smaller extent, do have a positive effect (compare Figure 87 and corresponding explanations).

With regard to the simulation results, concentrating mitigation on a continuously operating monitoring and early-warning system and eventually combining this passive measure with small, economic interventions, primarily drainage measures in key areas of the landslide, appears to be the most recommendable strategy. But as the model assumed a constant drawdown of the water level, the simulated positive effect of drainage systems can only be expected, if these systems are fully functional over many years. Therefore, according to the modelling results, any type of drainage systems eventually planned for the Corvara landslide must be designed in a way that they are able to tolerate a large amount of displacements, and regular maintenance and repair works must be assured, elsewise the interventions would be of little use.

The additional data gathered by a monitoring and early-warning system could be used directly for an improved calibration of the model at hand or to set up refined and more realistic models of source area S3 and also other landslide sectors. For this purpose, it would be important that monitoring be carried out on the long term and displacements of some key sectors be recorded in short intervals or continuously in order to capture the seasonal variations. When analysed on the long term together with the data of an on-site meteorological station, these time series would facilitate a better understanding of the hydrogeological context and the interrelationships of slope movement with different meteorological situations (compare Section 4.5.6., pages 139 to 142).

Applying the proposed Inverse Parameter Identification approach would assure that any additional reference data set leads to a quantifiable increase of calibration quality and of the determinateness of the modelled problem. This is an important advantage of the inverse modelling concept. Furthermore, before the installation of new measuring devices, the inverse approach permits to determine where, based on the current understanding of the slope that is reflected by the numerical model, additional measurements would be most effective, i.e. where these measurements would bring about the largest gain in determinateness of the problem. This so-called "marking of sensitive points" works similar to the analysis of the sensitivity of displacements at one specific point with respect to certain parameters, as is was described in Section 4.7.8. and Figure 82 (page 183). The longer and the more numerous time series of measurements are, the more parameters can be identified. In principle, the Inverse Parameter Identification Technique can be used also for identifying hydrogeological and geometrical parameters (as for example the depth of a shear-zone) or for analysing if and how well these parameters could be determined based on given reference data sets. With an increased reference data set, additional parameters could be identified, which until now had to be fixed based on assumptions. Quantitative results – deviation values, correlations etc. – obtained by the statistical analyses and during the optimisation allow for judging whether a refinement of the model improves its quality or if an apparently more realistic model is related to a decreased determinateness of the modelled problem. This type of information can hardly be gained when model calibration is performed by means of a traditional trial-and-error procedure.

4.9. Conclusions drawn with respect to the Corvara case study: requirements, difficulties and advantages of the presented numerical modelling concept

As it was demonstrated for the example of Corvara, the presented numerical modelling concept demands for an interdisciplinary approach, comprising Geology, Geomorphology, Geotechnics and Civil Engineering. In this kind of complex geotechnical problems on large and heterogeneous natural slopes, extensive investigation in the field – mapping of geology and geomorphology, geophysical surveying, subsurface investigation and groundwater monitoring – is required to obtain a realistic geometry model and to set appropriate hydrogeological boundary conditions. Comprehensive measurement series of surface and sub-surface displacements are needed as reference data for model calibration by means of inverse methods.

Anyway, the application of quite sophisticated numerical procedures generally calls for high-quality engineering-geological field data (e.g. *Stead et al. 2006*). The efforts and costs related to collecting these data are paying off in many ways, because the data are valuable also for risk mapping and for the design of early-warning systems. In the case of Corvara, many of the data taken as a basis for the presented numerical model were gathered in the course of hazard and risk assessment (*Corsini et al. 2005, Panizza et al. 2006*).

When applying the described inverse modelling approach, any additional measured value or series of measurements can be used directly for refining model calibration. Thereby determinateness of the inverse problem improves with each additional measured value. The great advantage of the presented approach is that this calibration quality can be quantified (via the objective function). Beyond that, this method provides statistical information about sensitivity and interdependence of model parameters.

Generally, "field problems" of the presented type are underdetermined. Therefore, the construct of ideas composing a model of such a problem must be tested incessantly for its plausibility, and remain revisable. Consequently, inverse modelling methods are convenient because they are relatively objective and traceable. For these reasons, their further development for geotechnical applications is expected to provide interesting possibilities in Engineering Geology.

5. Summary and outlook

This thesis dealt with natural slope movements of the earthflow-type, which are of continuous and long-lasting nature, i.e. taking place over years to thousands of years. These movements affect and are due to the presence and formation of thick soil covers (metres to tens of metres), covering in-situ rock masses. The overall phenomenon is characterised by pronounced stages of activity that are marked by the varying speed of the movements, which are relatively slow (ranging from mm/year to some tens of meters per day) compared to other landslide types.

The main thematic focus of this work was on the investigation of the so-called source areas located in the upper parts of the affected slopes. On the basis of two present-day and socio-economically relevant case studies in Northern Italy -- the Valoria earthflow in the Emilia Apennines and the Corvara earthflow in the Dolomites -- selected aspects of the source areas were investigated. This was done mainly by means of numerical modelling. The numerical modelling concept which was developed and applied in this work has the following four components: A geometry model that describes the positions and the course of the most important shear-zones of the studied slope and the geometry of the blocks and layers between these zones, an adequate material model that takes into consideration weights and stiffnesses of these blocks and layers, a creep model that describes the time-dependent material behaviour of the shear-zones, and finally, a hydrogeological model that describes the distribution of pore water pressures inside the slope, which is used as an important boundary condition. This concept can be put into practice with different numerical methods. Here, a continuum mechanical approach with the Finite-Element method and with the Soft Soil Creep model as constitutive model for the shear-zone materials was used.

In the case study of the Corvara earthflow, the transient distribution of pore water pressures inside the slope with time was modelled, and furthermore, the numerical modelling concept was combined with an inverse modelling strategy, calibrating the Finite-Element model against displacement rates measured in the field by iteratively varying the material-parameters of the shear-zones. This inverse modelling concept is based on statistical analyses and optimisation algorithms.

At the beginning of this thesis, the earthflow phenomenon was described and investigated in a general manner. Earthflows were studied with respect to their occurrence, morphological features, the properties of the materials involved, the hydrogeology of the affected slopes and the characteristics of the movements taking place. This was done based on a literature review, detailed field work carried out in the Lahnenwiesgraben testing area in the Bavarian Alps, Germany, and the classification of soil samples taken of different types of earthflow-prone materials at different sites in the alpine region.

All the earthflows studied in the Lahnenwiesgraben testing area were related to two stratigraphic units: Kössener Schichten of the Late Triassic, consisting of interbedded strata, made up of claystones, shales, marls and limestones; and Liasfleckenmergel (Allgäu formation), Early Jurassic, made up mainly of marly claystones and weakly-bound sandstones. Earthflow phenomena in this testing area were also found affecting moraine deposits, but in all cases, these moraine deposits consisted predominantly of weathering material of the Kössener Schichten strata.

The earthflows in the Lahnenwiesgraben valley mostly occupied depressions of bedrock topography which were characterised by an increased thickness of soil cover. Where bedrock topography differed from surface topography, the extension (and also the movement) of the earthflow bodies was oblique to the slope.

The analysis of the movements of the earthflows of the Lahnenwiesgraben valley in the context of this work lead to the conclusion that all the earthflows in this testing area are at present in the dormant stage. In this stage, their movements are caused by slow but ongoing creep deformations which occur mainly along shear-zones and take place under drained conditions. The speed of these slope movements, i.e. the rate of the creep deformations, is controlled by the level of pore water pressures in the slope. Variations of these pore water pressures – and consequently also of the displacement rates – are moderate and associated with seasonal changes in the amount of infiltration.

A series of soil samples were taken from materials typically encountered in connection with earthflow activity in the alpine realm. In this way, earthflow materials of different stratigraphic origin were sampled. The fraction with grain-size < 2 mm was classified in the laboratory, while the fraction with grain-size > 2 mm was classified according to a petrographic and lithological description of the rock fragments and rated according to the predominant shape, degree of rounding and degree of weathering. Results were compared with respect to the positions of the soil samples on the affected slope, such as the source area, track zone or accumulation area of the earthflow, as well as within the earthflow deposits, i.e. between shear-zones or at the depth of the presumed shear-zones.

Compared to the overall range of grain-size distributions of the actual earthflow materials, the samples taken at the depth of shear-zones were characterised by a relatively high proportion of clay-size particles and medium to coarse sand-size particles, while the amount of the silt and fine sand fraction was relatively low. On average, this group of samples also showed higher liquid limits and plasticity indexes.

Results obtained by the semi-quantitative rating procedure based on the petrographic and lithological description of the fraction > 2 mm were quite similar for most of the samples. The majority of the samples contained rock-types which exhibited a noticeably to extremely low slaking durability. These observations suggested that lithologies of low slaking durability are one of the main geological causes for earthflows, as they favour the fast and deep weathering processes that lead to the formation of the thick soil covers required for the development of these landslides.

The source area of the frequently reactivated Valoria earthflow was chosen to study the medium-term history (years to decades) of the slope morphology in the crown and head zone of this landslide by means of a qualitative reconstruction in a numerical simulation. Using this tool, different phases of activity and critical events were analysed together with a large amount of different field- and monitoring-data as well as relatively precise topographic data, which document the slope morphology before and after the reactivation events of 2001, 2005 and 2007. This was done to obtain an improved geological, kinematical and mechanical understanding of the slope history. To account for different possible interpretations of the field- and monitoring-data with respect to the subsurface geometry, two different 2D geometry models were set up along a representative section through the upper source area, resulting in two different numerical models -- the first of which was simplified and more robust, and the second of which was more complex and detailed but implied a higher number of assumptions. The different monitored displacement rates observed in different parts of the studied slope section during the dormant, suspended and preparatory phases were reproduced qualitatively by the models. Also, the initiation of the critical (active) phases and the rough spatial distribution of the displacements were explained logically by the simulations. The results suggest that the principal engineering-geological hypotheses taken as a basis for the numerical models are plausible. According to these hypotheses, the slope evolution is the result of a complex interaction between changing material properties, changing hydrological conditions and deformations. The different activity phases of the Valoria earthflow are linked not only to climatically-driven changes of the hydrological situation, but also to the deterioration of material properties in the source area that is taking place with time and as a consequence of the deformations and related changes of the slope geometry.

The results of the Valoria case study lead to the conclusion that the deterioration of the material properties is a consequence of the geological composition and the tectonic history of the involved rock masses, and therefore, will not stop. Further deformations, which again have a negative impact on the material properties, and further partial or total reactivations of the earthflow, must always be expected. Passive mitigation measures are indispensable regardless of eventual active mitigation measures. Also in the future, mitigation measures must include a continuous monitoring of the movements and of the hydrogeological conditions to recognise new critical events in time.

In the second case study, an important part of the source area of the Corvara earthflow (source area S3) was selected to analyse seasonal changes of monitored movement rates during a relatively calm phase of the landslide's history together with on-site meteorological data. A 2D numerical model based on simplifying geotechnical and hydrogeological assumptions, which were abstracted from a large amount of geological, hydrogeological, geomorphological, geophysical and geotechnical field data, was set up. Inverse modelling methods were used to calibrate this numerical model and to back-calculate important material parameters on the basis of field-measurements. The inverse modelling strategy was further utilized to generate statistical information on model behaviour, on the sensitivity of model parameters and on the quality of the obtained calibration.

Simulation and back-analysis of laboratory tests – oedometer tests, an isotropic compression test and a deviatoric creep test – showed that the constitutive model (the Soft Soil Creep model) correctly simulated the material behaviour of the fine-grained soil-matrix governing the behaviour of the landslide material, and that the Inverse Parameter Identification Technique identified parameter sets that allowed for an approximate reproduction of this behaviour under the boundary conditions and in the stress ranges of the experiments.

The comparison of displacement vectors calculated by the calibrated numerical slope model of source area S3 of the Corvara earthflow for the three years of the monitoring period between 2001 and 2004 with those determined by means of GPS (Global Positioning System) measurements gave the following results. Reproduction of the horizontal displacements was very good. Except for two GPS points, vertical displacements were overestimated. However, the rough direction of all simulated displacements was sub- parallel to the slope as it was generally observed in measurements in all sectors of the Corvara landslide. Qualitatively, the modelled displacement profile in the lower part of the slope section was in accordance with that measured by the nearby inclinometer.

When comparing the measured time series of the horizontal displacements at the GPS points in source area S3 with the simulation results, the following observations were made. For one group of measurement points, the simulated curves were close to the real time series, except for small differences. For another group of points the total horizontal displacements were reproduced approximately by the model, but displacements were overestimated in the phases of low water pressures and underestimated in the phases of high water pressures. This result is attributed to the fact that the modelling of the pore pressures was based on simplifying hydrogeological assumptions. In the case of the first group of points, these assumptions were apparently more appropriate for describing reality as in the case of the second group of points. The measurement points belonging to the first group were located in that slope region where most of the hydrogeological data could be collected.

The calibrated numerical model of source area S3 of the Corvara earthflow could be used successfully in a civil engineering study (that was based on the results of this thesis) to implement different combinations of slope reinforcement and drainage measures and to pre-evaluate their effects.

Generally, the results obtained in both case studies demonstrate the importance of detailed field investigations and long-term measurements as basis for numerical models. The determinateness of the modelled problem depends on the availability of such data, which is limited by the large dimensions of the investigated phenomena and by the heterogeneity of the involved materials as well as the uncertainties underlying the determination of the geometry of these problems. In the future, new and more economic measurement techniques might bring about improvements. The examples of Valoria and Corvara show that short-term success of planned active mitigation measures is unlikely, although in the case of Corvara, there are some indicators that drainage measures might have positive influence on the longer term. In both cases, passive mitigation measures, especially monitoring of slope movements and hydrogeology, are therefore most important. The concept presented together with the Corvara case study, i.e. the combination of the forward numerical model with an inverse modelling strategy, allows direct application of these measurements for improved model calibration and refinement. The great advantage of the presented approach is that calibration quality can be quantified (via an objective function). Beyond that, the method provides statistical information about sensitivity and interdependence of model parameters. Quantitative results – deviation values, correlations etc. – obtained by the statistical analyses and during the optimisation procedure provide feedback on the refinement of the model and whether an apparently more realistic model is related to a decreased determinateness of the modelled problem.

With an increased reference data set (i.e. measured data of any type), it would be possible to identify additional parameters, which until now had to be fixed based on assumptions. In principle, any additional value measured in the field brings about a quantifiable improvement of the overall model in this sense. Due to the characteristics of the earthflow-type landslides studied in this work, an increase of the temporal resolution of the displacement measurements and of the spatial resolution of the pore pressure measurements are most promising. In further research, the inverse modelling strategy may also be used to test these hypotheses and clarify which specific type of measurement data at which position along or within the slope would be most efficient for increasing the determinateness of the overall modelled problem.

Zusammenfassung

Gegenstand der vorliegenden Arbeit sind natürliche Hangbewegungen vom Typ eines Schuttstroms, die kontinuierlich und langandauernd (im Zeitraum von Jahren bis zu Tausenden von Jahren) stattfinden. Hangbewegungen dieser Art kommen in und aufgrund von mächtigen Lockergesteinsdecken vor, die zwischen einigen Metern und mehreren 10er Metern mächtig sein können und auf anstehendem Fels liegen. Schuttströme haben Phasen unterschiedlich ausgeprägter Aktivität. Im Vergleich zu anderen Arten von Hangbewegungen bewegen sie sich mit Geschwindigkeiten von einigen Millimetern pro Jahr bis zu mehreren 10er Metern pro Tag relativ langsam.

Das Hauptaugenmerk der Arbeit liegt auf der Untersuchung der sogenannten Liefergebiete, die sich in den oberen Bereichen der betroffenen Hänge befinden. Anhand von zwei aktuellen Fallstudien in Norditalien, die auch sozioökonomisch relevant sind – dem Schuttstrom von Valoria im Apennin der Emilia und dem Schuttstrom von Corvara in den Dolomiten – wurden ausgewählte Aspekte der Liefergebiete untersucht. Dies wurde vor allem anhand von numerischen Modellen durchgeführt.

Das Konzept zur numerischen Modellierung, das in dieser Arbeit entwickelt und angewendet wurde, besteht aus vier Komponenten:

1. einem Geometriemodell, das die Lage und den Verlauf der wichtigsten Scherzonen des untersuchten Hanges beschreibt, sowie die Geometrie der Blöcke und Schichten zwischen diesen Scherzonen;

2. einem geeigneten Materialmodell, das die Gewichte und Steifigkeiten dieser Blöcke und Schichten beschreibt;

3. einem Kriechmodell, das das zeitabhängige Materialverhalten der Scherzonen beschreibt und

4. einem hydrogeologischen Modell, das die Verteilung der Porenwasserdrücke im Hang beschreibt, was als eine wichtige Randbedingung verwendet wird.

Dieses Konzept kann durch verschiedene Methoden in die Praxis umgesetzt werden. In der vorliegenden Arbeit wurde ein kontinuumsmechanischer Ansatz gewählt, mit der Finite-Elemente-Methode und mit dem Soft-Soil-Creep Modell als konstitutivem Modell für die Materialien der Scherzonen.

In der Fallstudie des Schuttstroms von Corvara wurde auch die zeitabhängige Entwicklung der Porenwasserdruckverteilung im Hang modelliert und darüber hinaus wurde das numerische Modellierungskonzept mit einer inversen Modellierungsstrategie kombiniert. Dabei wurde das Finite-Elemente-Modell anhand von im Gelände gemessenen Verschiebungsraten kalibriert, indem die Materialparameter der Scherzonen iterativ variiert wurden. Dieses Konzept zur inversen Modellierung basiert auf statistischen Auswertungen und Optimierungsalgorithmen.

Am Anfang dieser Dissertationsschrift wurde das Phänomen der Schuttströme im allgemeinen behandelt.

Durch Literaturstudien, detaillierte Feldarbeit im Testgebiet Lahnenwiesgraben in den Bayerischen Alpen (Deutschland) und die Klassifizierung von Bodenproben verschiedener Materialien, die zur Schuttstrombildung neigen, wurden derartige Hangbewegungen hinsichtlich

- ihrer Verbreitung
- ihrer morphologischen Besonderheiten
- der Eigenschaften der an diesem Phänomen maßgeblich beteiligten Materialien
- der Hydrogeologie der betroffenen Hänge und
- der Merkmale der stattfindenden Bewegungen untersucht.

Alle im Testgebiet Lahnenwiesgraben untersuchten Schuttströme standen mit den stratigraphischen Einheiten Kössener Schichten (der Obertrias) und Liasfleckenmergel (Allgäu Schichten, aus dem unteren Jura) in Verbindung. Letztgenannte bestehen hauptsächlich aus Ton- und schwachgebundenen Sandsteinen. Die Kössener Schichten sind aus einer Wechsellagerung von Tonsteinen, Schiefertonen, Mergeln und Kalksteinen zusammengesetzt.

Es wurde auch beobachtet, dass Moränenablagerungen betroffen waren. Diese bestanden überwiegend aus Verwitterungsmaterial der Kössener Schichten.

Die Schuttströme im Lahnenwiesgraben nahmen - im Hinblick auf die Topographie der darunterliegenden Felsoberfläche - meist Hangmulden ein. Jene zeichneten sich durch eine erhöhte Mächtigkeit der Lockergesteinsdecke aus. Wo die Oberfläche des anstehenden Gebirges von der Topographie des Hanges abwich, konnte beobachtet werden, dass die Ausrichtung der Schuttstromkörper und auch deren Bewegungsrichtung schräg zum Hang verlief.

Aus den in dieser Arbeit untersuchten Bewegungen der Schuttströme im Tal des Lahnenwiesgrabens lassen sich folgende Schlussfolgerungen ableiten: Alle Schuttströme dieses Testgebietes befanden sich im Untersuchungszeitraum in der Ruhephase. In diesem Stadium sind ihre Bewegungen auf langsame aber andauernde Kriechverformungen zurückzuführen, die hauptsächlich entlang von Scherzonen und unter drainierten Bedingungen stattfinden. Die Geschwindigkeit dieser Bewegungen, die Kriechrate, wird durch den Porenwasserdruckspiegel im Hang gesteuert. Die Schwankungen der Porenwasserdrücke und folglich auch die Schwankungen der Verschiebungsraten sind moderat und hängen mit saisonalen Schwankungen der Menge der Infiltration zusammen.

Von Materialien, die für die Schuttstromaktivität im Alpenraum charakteristisch sind, wurde eine Reihe von Bodenproben genommen und dabei Schuttstrommaterialien unterschiedlicher stratigraphischer Herkunft untersucht. Die Korngrößenfraktion < 2 mm wurde im Labor klassifiziert.

Für die Korngrößenfraktion > 2 mm wurde eine petrographische und lithologische Beschreibung der Gesteinsbruchstücke sowie eine Einordnung und Bewertung nach der vorherrschenden Kornform und dem Rundungs- und Verwitterungsgrad durchgeführt. Die Ergebnisse wurden hinsichtlich der Positionen der Proben auf dem Hang, wie z. B. Liefergebiet, Transportkanal und Akkumulationszone des Schuttstromes, verglichen. Weiterhin wurden Vergleiche innerhalb der Schuttstromkörper angestellt, d. h. zwischen Scherzonen oder in der Tiefenlage der vermuteten Scherzone.

Im Vergleich zum Gesamtbereich der Kornverteilungen der aktuellen Schuttstrommaterialien zeichneten sich die Proben, die in der Tiefenlage der Scherzonen genommen wurden, durch einen relativ hohen Anteil an tonigem Material sowie gleichzeitig der Mittel- und Grobsandfraktion aus, während die Menge der Silt- und Feinsandfraktion in diesen Proben relativ niedrig war. Im Durchschnitt wies diese Gruppe von Proben auch höhere Fließgrenzen und Plastizitätszahlen auf. Die Ergebnisse der semiquantitativen Untersuchung basierend auf der petrographischen und lithologischen Beschreibung der Gesteinsbruchstücke der Korngrößenfraktion > 2 mm waren für die Mehrzahl der Proben sehr ähnlich. Die Proben enthielten vorwiegend Gesteinsarten, die sich durch eine bemerkenswerte bis extreme veränderliche Festigkeit auszeichneten. Diese Beobachtung legte nahe, dass Gesteinsarten veränderlicher Festigkeit eine der wichtigsten geologischen Ursachen für Schuttströme sind. Denn sie fördern die schnellen und tiefgründigen Verwitterungsprozesse, die zur Entstehung der mächtigen Lockergesteinsdecken führen, die wiederum für die Entwicklung von solchen Hangbewegungen verantwortlich sind.

Das Liefergebiet des häufig reaktivierten Schuttstroms von Valoria wurde ausgewählt, um die mittelfristige (im Zeitraum von Jahren bis Jahrzehnten) Entwicklung der Hangmorphologie im Kronenbereich und im Kopfbereich dieser Hangbewegung mittels einer qualitativen Rekonstruktion in einer numerischen Simulation zu untersuchen. Unter Verwendung dieser Methode wurden verschiedene Aktivitätsphasen der Hangbewegung und kritische Ereignisse zusammen mit einer großen Menge an Feld- und Messdaten sowie recht genauer topographischer Daten, die die Hangmorphologie jeweils vor und nach den Reaktivierungsereignissen von 2001, 2005 und 2007 genau dokumentieren, im Modell nachgestellt.

Zweck dieser Untersuchung war es, ein vertieftes geologisches, kinematisches und mechanisches Verständnis der Hangentwicklung zu erhalten. Um unterschiedlichen möglichen Interpretationen der Feld- und Messdaten hinsichtlich der unterirdischen Geometrie des Hanges Rechnung zu tragen, wurden zwei verschiedene zweidimensionale Geometriemodelle entlang eines repräsentativen Profilschnittes durch das obere Liefergebiet entwickelt: erstens, ein stärker vereinfachendes und robusteres Modell und zweitens, ein komplexeres und detaillierteres Modell, das eine größere Anzahl Modellannahmen benötigte.

Die unterschiedlichen gemessenen Verschiebungsraten, die in verschiedenen Bereichen des Hangprofils beobachtet wurden – Phasen relativer Ruhe (dormant phases), Phasen kürzlich eingestellter Aktivität (suspended phases) sowie Vorlaufphasen größerer Ereignisse (preparatory phases) – wurden in der Simulation am Computer nachgebildet. Auch der Beginn der kritischen Phasen (active/critical phases) und die grobe räumliche Verteilung der Verschiebungen zu diesem Zeitpunkt konnten durch die Simulation erklärt werden.

Die Ergebnisse legen nahe, dass die prinzipiellen ingenieurgeologischen Annahmen, die den Modellen zugrundegelegt wurden, plausibel sind. Nach diesen Hypothesen ist die Hangbewegung das Ergebnis einer komplexen Wechselwirkung zwischen sich verändernden Materialeigenschaften, schwankenden hydrologischen Verhältnissen und zwischen den Verformungen, die im Hang stattfinden. Die verschiedenen Aktivitätsphasen des Schuttstroms von Valoria stehen nicht allein im Zusammenhang mit klimatisch gesteuerten Änderungen der hydrologischen Situation, sondern auch mit den sich verschlechternden Materialeigenschaften des Gebirges im Liefergebiet des Schuttstromes mit der Zeit und als Folge der Hangdeformationen und der damit verbundenen Änderungen der Geometrie des Hanges.

Die Ergebnisse der Fallstudie von Valoria legten nahe, dass die Verschlechterung der Materialeigenschaften in der geologischen Zusammensetzung und tektonischen Vorgeschichte des betroffenen Gebirges begründet ist und sich deshalb fortsetzen wird: Weitere Verformungen, die wiederum einen negativen Effekt auf die Materialeigenschaften haben sowie weitere teilweise oder vollständige Reaktivierungen des Schuttstromes müssen jederzeit erwartet werden. Passive Maßnahmen zur Minderung des Risikos, das von dieser Hangbewegung ausgeht, sind unabdingbar, unabhängig davon, ob aktive Sanierungsmaßnahmen in Erwägung gezogen werden. Auch in der Zukunft müssen die Maßnahmen, die zur Verminderung des Risikos unternommen werden, vorrangig eine ständige Überwachung der Bewegungen und der hydrogeologischen Verhältnisse beinhalten, um neue kritische Ereignisse (active phases) rechtzeitig erkennen zu können.

In der zweiten Fallstudie wurde ein wichtiger Bereich des Liefergebietes des Schuttstromes von Corvara (Liefergebiet S3) ausgewählt, um die saisonalen Schwankungen der im Gelände gemessenen Bewegungsraten in einer Phase relativer Ruhe in der Geschichte dieses Schuttstromes im Zusammenhang mit vor Ort aufgenommenen meteorologischen Daten zu analysieren. Ein zweidimensionales numerisches Modell, basierend auf vereinfachenden geotechnischen und hydrogeologischen Annahmen wurde erstellt. Die Modellannahmen wurden aus der Auswertung einer großen Menge geologischer, hydrogeologischer, geomorphologischer, geophysikalischer und geotechnischer Daten getroffen. Um dieses numerische Modell anhand von Feldmessungen zu kalibrieren und um wichtige Materialparameter rückzurechnen, kamen inverse Modellierungstechniken zur Anwendung. Die inverse Modellierungsstrategie wurde weiterhin dazu verwendet, statistische Informationen über die Sensitivität von Modellparametern und über die Qualität der erzielten Modellkalibrierung, zu gewinnen.

Simulation und Rückwärtsrechnung von Laborversuchen – Oedometerversuche, ein isotroper Kompressionsversuch und ein deviatorischer Kriechversuch – zeigten, dass das verwendete Stoffmodell (Soft-Soil-Creep-Modell) das Materialverhalten der feinkörnigen Matrix, die das Verhalten des Schuttstrommaterials maßgeblich bestimmt, korrekt abbildet. Weiterhin lässt sich feststellen, dass die inverse Methode zur Parameteridentifikation Parametersätze lieferte, die eine näherungsweise Wiedergabe des Materialverhaltens unter den Randbedingungen und im Spannungsbereich der Experimente erlaubten. Der Vergleich der Verschiebungsvektoren, die durch das numerische Hangmodell vom Liefergebiet S3 des Schuttstromes von Corvara für den Messzeitraum von 2001 bis 2004 berechnet wurden, mit denen, die mittels GPS (Global Positioning System)- Messungen bestimmt wurden, erbrachte das folgende Ergebnis: Die Gesamtmenge der horizontalen Verschiebungen wurde sehr gut wiedergegeben. Mit Ausnahme von zwei Messpunkten wurden die vertikalen Verschiebungen durch das Modell überschätzt. Trotzdem war jedoch die grobe Richtung ungefähr parallel zur Hangneigung wie es grundsätzlich in den Messungen in allen Bereichen der Hangbewegung von Corvara beobachtet wurde. Qualitativ stimmte das simulierte Verschiebungsprofil im unteren Bereich des Hanges mit dem im nahegelegenen Inklinometer gemessenen überein. Anhand des Vergleiches der gemessenen Zeitreihen der horizontalen Verschiebungen an den GPS-Messpunkten im Liefergebiet S3 mit den Ergebnissen der Simulation ergaben sich folgende Beobachtungen: Bei einer Gruppe von Messpunkten lagen die simulierten Kurvenverläufe bis auf kleine Unterschiede nahe an den tatsächlichen Zeitreihen. Bei der anderen Gruppe von Messpunkten wurden zwar die Gesamtverschiebungen näherungsweise durch das Modell wiedergegeben, jedoch die Verschiebungen in den Phasen niedriger Porenwasserdrücke über- und in den Phasen hoher Wasserdrücke unterschätzt. Dieses Ergebnis lässt sich damit begründen, dass die Modellierung der Porenwasserdrücke auf vereinfachenden hydrogeologischen Annahmen basieren musste. Im Fall der ersten Gruppe von Messpunkten waren diese Annahmen offenbar besser geeignet, die Realität zu beschreiben, als im Fall der zweiten Gruppe von Messpunkten. Die Messpunkte der ersten Gruppe lagen in dem Hangbereich, für den die meisten hydrogeologischen Daten gesammelt werden konnten.

Das vorgestellte kalibrierte numerische Modell des Liefergebietes S3 konnte inzwischen bereits in einer Forschungsarbeit im Bauingenieurwesen, die auf den Ergebnissen der vorliegenden Dissertationsschrift aufbaute, erfolgreich weiterverwendet werden: verschiedene Kombinationen von aktiven baulichen Sanierungsmaßnahmen (Verbauwände, Anker, Drainagen) wurden in das numerische Modell implementiert, um deren Auswirkungen zu beurteilen.

Ausblick

Generell stellen die Ergebnisse beider Fallstudien, Valoria und Corvara, die Bedeutung detaillierter Felduntersuchungen und Langzeitmessungen als Grundlage für numerische Modelle heraus. Die Bestimmtheit des modellierten Problems hängt von der Verfügbarkeit solcher Daten ab, die wiederum durch die großen Dimensionen der untersuchten Phänomene und durch die Inhomogenität der beteiligten Materialien sowie die Unsicherheiten, denen die Bestimmung der Geometrie dieser Probleme unterliegt, begrenzt wird. Künftig könnten neue und kostengünstigere Messtechniken in diesen Bereichen Verbesserungen bewirken.

Die untersuchten Beispiele der Hänge von Valoria und Corvara legen nahe, dass ein kurzfristiger Erfolg baulicher Sanierungsmaßnahmen unwahrscheinlich ist. Nichtsdestotrotz ließen sich im Fall von Corvara Hinweise darauf finden, dass Dränagemaßnahmen langfristig durchaus einen positiven Einfluss haben könnten. In beiden Fallstudien sind deswegen passive Maßnahmen zur Verminderung des Risikos, vor allem die messtechnische Überwachung der Hangbewegungen und der Hydrogeologie, von größter Bedeutung. Das im Zusammenhang mit der Fallstudie von Corvara angewendete Konzept – die Kombination des numerischen Modells der Vorwärtsrechnung mit einer inversen Modellierungsstrategie – ermöglicht es, diese Messungen direkt für eine verbesserte Modellkalibrierung und eine Verfeinerung des Modells zu nutzen. Der große Vorteil dieses Ansatzes besteht darin, dass die Qualität der Modellkalibrierung mittels einer Zielfunktion quantifiziert werden kann. Darüber hinaus stellt die Methode statistische Informationen über Sensitivität und wechselseitige Abhängigkeit von Modellparametern zur Verfügung. Quantitative Ergebnisse (Abweichungswerte, Korrelationen usw.), die durch die statistischen Auswertungen und während des Optimierungsverfahrens gewonnen werden, ermöglichen es zudem zu beurteilen, ob eine Verfeinerung des Modells eine tatsächliche Verbesserung bringt, oder ob ein scheinbar realistischeres Modell mit einer niedrigeren Bestimmtheit des modellierten Problems insgesamt verbunden ist. Mit einem größeren Referenz-Datensatz (d. h. Messdaten jeglichen Typs) wäre es möglich, zusätzliche Parameter zu identifizieren, die bis jetzt lediglich aufgrund von Annahmen festgelegt werden mussten. Grundsätzlich bringt jeder zusätzliche Messwert eine quantifizierbare Verbesserung des Gesamtmodells in diesem Sinne.

Aufgrund der Eigenschaften der in der vorliegenden Promotionsschrift untersuchten Hangbewegungen vom Typ des Schuttstromes wären eine Erhöhung der zeitlichen Auflösung der Bewegungsmessungen und eine Erhöhung der räumlichen Auflösung der Porenwasserdruckmessungen am vielversprechendsten. In einer weitergehenden Untersuchung könnte die inverse Modellierungsstrategie auch dazu verwendet werden, solche Hypothesen zu testen und herauszufinden, welche Art von Messdaten an welcher Stelle entlang des Hanges oder innerhalb desselben am effizientesten wären, um die Bestimmtheit des Gesamtproblems zu erhöhen.

6. Literature references

Agliardi, F., Crosta, G., Zanchi, A., 2001, Structural constraints on deep-seated slope deformation kinematics. Eng. Geol. 59 (1/2), 83–102

Aleotti, P. and Chowdhury, R., 1999, Landslide hazard assessment: summary review and new perspectives. – Bull. Engng. Geol. Environ., 58, 21-44

Angeli, M.-G., Menotti, R. M., Pasuto, A., Silvano, S., 1991, Landslides studies in the eastern Dolomites Mountains, Italy, Landslides, Bell (ed.), 275-282, 1991 Balkema, Rotterdam

Angeli, M.-G., Gasparetto, P., Pasuto, A., Silvano, S., Menotti, R. .M.,. 1996 A, Examples of mudslides on low-gradient clayey solpes. Proc. 7th Int. Symp. Landslides, Trondheim 1: 141 – 145. Rotterdam, Balkema

Angeli, M.-G., Gasparetto P., Menotti, R. M., Pasuto, A., Silvano, S., 1996 B, A visco-plastic model for slope analysis applied to a mudslide in Cortina d'Ampezzo, Italy, Quarterly Journal of Engineering Geology, 29, 233-240, The Geological Society

Angeli, M.- G., Buma, J., Gasparetto, P., Pasuto, A., 1998, A combined hillslope hydrology/stability model for low-gradient clay slopes in the Italien Dolomites, Engeneering Geology 49 (1998) 1-13, Elsevier

Avolio, M. V., Di Gregorio, S., Mantovani, F., Pasuto, A., Rongo R., Silvano, S., Spataro, W., 2000, Simulation of the 1992 Tessina landslide by a cellular automata model and future hazard scenarios, JAG, Vol 2, Issue 1, 2000

Baldi, A. M., De Luca, J., Lucente, C. C., Sartini, G., 2009, Indagine di sismica a rifrazione per lo studio della frana attiva dei Boschi di Valoria, Atti del 3° Congresso Nazionale AIGA – Centro di GeoTecnologie, Università degli Studi di Siena, San Giovanni Valdarno (AR), 25-27 Febbraio 2009

Bammer, O., 1984, Massenbewegungen im Raume Bad Goisern, Oberösterreich, Internat. Symp. Interpraevent 1984, Villach, 3: 167-180, Villach

Baumgartner, P., 1981, Erd- und Schuttströme im Gschliefgraben bei Gmunden am Traunsee (O.Ö.). Zur Geologie, Entstehung, Entwicklung und Sanierung. Mitt. Ges. Geol. Bergbaustud. Österr. 27: 19 – 38. Wien

Baumgartner, P. and Mostler, H. 1978, Zur Entstehung von Erd- und Schuttströmen am Beispiel des Gschliefgrabens bei Gmunden (Oberösterreich), Geol. Palaont. Mitt. Innsbruck, Bd. 8, Festschrift W. Heissel, S. 113-122, Innsbruck, Sept. 1978

Baron, I., Agliardi, F., Ambrosi, C., Crosta, G. B., 2005, Numerical analysis of deep-seated mass movements in the Magura Nappe; Flysch Belt of the Western Carpathians (Czech Republic). Natural Hazards and Earth System Sciences, 5, 367–374

Berti, M., Genevois, R., Ghirotti, M., Tecca, P. R., 1994, The use of drainage systems in the stabilization of a complex landslide, 7th International IAEG Congress, 1994 Balkema, Rotterdam

Berti, M. and Simoni, S., 2009, in press, Field evidence of pore pressure diffusion in clayey soils prone to landsliding. Journal of Geophysical Research

Bertini, T., Cugusi, F., D'Elia, B., Lanzo, G., Rossi-Doria, M., 1991, Slow movement investigations in clay slopes, Landslides, Bell (ed.), 1991 Balkema, Rotterdam

Bertolini, G., 2003, Frane e variazioni climatiche nell'Appennino emiliano, Tesi di Dottorato in Geologia dell' Ambiente, Dipartimento di Scienze della Terra, Università degli Studi di Modena e Reggio Emilia, 157 pp.

Bertolini, G. and Gorgoni, C., 2001, La Lavina di Roncovetro (Vedriano, Comune di Canossa, Provincia di Reggio Emilia), Quaderni di Geologia Applicata, 8 – 2 (2001)

Bertolini, G. and Pellegrini, M., 2001, The landslides of the Emilia Apennines (northern Italy) with reference to those which resumed activity in the 1994-1999 period and required Civil Protection interventions, in: Bertolini G., Pellegrini M., Tosatti G. (editors), Le frane della Regione Emilia-Romagna, oggetto di interventi di Protezione Civile nel periodo 1994-1999, Quaderni di Geologia Applicata 8(1):27-74, Pitagora Editrice, Bologna

Bertolini, G., Casagli, N., Ermini, L., Malaguti, C., 2004, Radiocarbon Data on Lateglacial and Holocene Landslides in the Northern Apennines, Natural Hazards 31: 645–662

Bertolini, G., Guida, M., Pizziolo, M., 2005, Landslides in Emilia Romagna region (Italy): strategies for hazard assessment and risk management. Landslides, 2: 302-312

Bertolini, G., Pellegrini M., Tosatti G. (editors), 2001, Le frane della Regione Emilia-Romagna, oggetto di interventi di Protezione Civile nel periodo 1994-1999, Quaderni di Geologia Applicata 8(1 and 2), Pitagora Editrice, Bologna

Bettelli, G. and De Nardo, M. T., 2001, Geological outlines of the Emilia Apennines (Italy) and introduction to the rock units cropping out in the areas of landslides reactivated in the 1994-1999 period., in: Bertolini G., Pellegrini M., Tosatti G. (editors), Le frane della Regione Emilia-Romagna, oggetto di interventi di Protezione Civile nel periodo 1994-1999, Quaderni di Geologia Applicata, 8(1), Pitagora Editrice, Bologna

Biavati, G. and Simoni, A., 2006, Analisi numeriche per la valutazione dell'efficacia di sistemi drenanti in versanti argillosi instabili, Numerical analysis for the evaluation of subsurface drains effectiveness, implications for the stability of clay-rich slopes, Giornale di Geologia Applicata 4 (2006) 101-108

Bockelmann, I., 1995, Der Schuttstrom von Maria Eck /Oberbayern, Diplomarbeit an der Friedrich-Alexander-Universität Erlangen-Nürnberg

Bonnard, C., Noverraz, F., Lateltin, O., Raetzo, H., 1995, Large Landslides and Possibilities of Sudden Reactivation. Felsbau 13 (6): 400 – 407. Essen: Glückauf

Bonomi, T. and Cavallin, A., 1999, Three-dimensional hydrological modelling application to the Alverà mudslide (Cortina d'Ampezzo, Italy). Geomorphology, 30, 189-199

Bonzanigo, L., Eberhardt, E., Loew, S., 2001, Hydromechanical factors controlling the creeping Campo Vallemaggia landslide. In UEF International Conference on Landslides - Causes, Impacts and Countermeasures, Davos, Switzerland, 17-21 June 2001, Edited by M. Kühne, H.H. Einstein, E. Krauter, H. Klapperich and R. Pöttler, Verlag Glückauf GmbH, Essen, pages 13-22

Borgatti, L., 2004, Relazioni tra frane e variazioni climatiche: concetti, metodi e casi di studio nelle Dolomiti, tesi per il conseguimento del titolo di dottore di ricerca, Università degli Studi di Modena e Reggio Emilia Dipartimento di Scienze della Terra, Italy

Borgatti, L. and Soldati, M., 2009, Landslides as a geomorphological proxy for climate change: A record from the Dolomites (Northern Italy), Geomorphology, In Press, DOI: 10.1016/j.geomorph.2009.09.015

Borgatti, L., Corsini, A., Barbieri, M., Sartini, G., Truffelli, G., Caputo, G., Puglisi, C., 2006, Large reactivated landslides in weak rock masses: a case study from the Northern Apennines (Italy). Landslides 3: 115-124

Borgatti, L., Ravazzi, C., Donegana, M., Corsini, A., Marchetti, M., Soldati, M., 2007 A,, A lacustrine record of early Holocene watershed events and vegetation history, Corvara in Badia, Dolomites (Italy). Journal of Quaternary Science, 22 (2): 173-189

Borgatti, L., Cervi, F., Corsini, A., Ronchetti, F., Pellegrini, M., 2007 B, Hydro-mechanical mechanisms of landslide reactivation in heterogeneous rock masses of the Northern Apennines (Italy), AEG Special Publication, 23, 749-758, 2007

Borgatti, L., Corsini, A., Marcato, G., Pasuto, A., Silvano, S., Zabuski, L., 2007 C, Numerical Analysis of Countermeasure Works influence on Earth Slide Stabilisation: A Case Study in South Tyrol (Italy), 1st North American Landslide Conference Landslides and Society: Integrated Science, Engineering Management and Mitigation, Proc. Intern. Conf. Vail, 3-8 June 2007, USA: AEG 23: 1007-1015

Borgatti, L., Corsini, A., Marcato, G., Ronchetti, F., Zabuski, L., 2008, Appraise the structural mitigation of landslide risk via numerical modelling: a case study from the northern Apennines (Italy), GEORISK, Volume 2, Issue 3, September 2008, Pages 141-160, published by Taylor & Francis

Bovis, M. J., 1985, Earthflows in the Interior Plateau, southwest British Columbia. Can. Geotechn. Journ. 1 (22): 313-334

Bovis, M. J., 1986, The morphology and mechanics of lage-scale slope movement, with particular reference to souethwest British Columbia, in: Abrahams AD (ed.) Hillslope processes. Allen & Unwin, Boston, pp. 319-341

Brideau, A., Yan, M., Stead, D., 2009, The role of tectonic damage and brittle rock fracture in the development of large rock slope failures, Geomorphology, Volume 103, Issue 1, Dating, triggering, modelling, and hazard assessment of large landslides, 1 January 2009, Pages 30-49

Bromhead, E. N., 1996, Slope stability modeling: an overview, in: Dikau, R., Brunsden, D., Schrott, L., Ibsen M.-L., Eds., Landslide recognition. – Wiley & Sons, Chichester, 231-235

Brunsden, D., 1999, Some geomorphological considerations for the future development of landslide models. – Geomorphology, 30, 13-24

Büch, F., 2003, Geologisch- geotechnische Untersuchungen gravitativer Massenbewegungen im Lahnenwiesgraben (Garmisch-Partenkirchen, Oberbayern), Diplomarbeit, Institut für Geologie und Mineralogie, Lehrstuhl für Angewandte Geologie, Friedrich-Alexander-Universität Erlangen-Nürnberg, 2003

Bunza, G., 1978, Bewegungsablauf und Sanierungsmöglichkeiten von Erdströmen. Geol. Pal. Mitt. Innsbruck 8 (Festschrift W. Heissel): 209 - 225. Innsbruck

Calvello, M. and Finno, R. J., 2004, Selecting parameters to optimize in model calibration by inverse analysis, Computers and Geotechnics, Vol. 31, pp. 411-425 (2004)

Cancelli, A., Crosta, G., Nardi, R., Pochini, A., 1991, An example of combined geological and geotechnical investigations for a landslide in seismic area, Landslides, Bell (ed.) 1991 Balkema, Rotterdam, pp. 1117-1122

Caris, J. P. T. and van Asch Th. W. J., 1991, Geophysical, geotechnical and hydrological investigations of a small landslide in the French Alps. Engineering Geology, Volume 31, Issues 3-4, 249-276

Caroli, N., 2009, Elaborazione di misure GPS riferite alla frana di Corvara in badia nel periodo 2001 – 2008, Tesi di Laurea in Scienze Geologiche, Facoltà di Scienze Matematiche Fisiche e Naturali, Università degli Studi di Modena e Reggio Emilia, Anno Accademico 2008/2009. Available as open file report at the Geological Office (Amt für Geologie und Baustoffprüfung) of the Autonomous Province of Bolzano/Bozen, Italy.

Carrera, J., Alcolea, A., Medina, A., Hidalgo, J., Slooten, L., 2005, Inverse Problem in Hydrogeology, Hydrogeological Journal, Vol. 13, pp. 206-222 (2005), Springer

Casagli, N., Dapporto, S., Ibsen, M. L., Tofani, V., Vannocci, P., 2006, Analysis of the landslide triggering mechanism during the storm of 20th–21st November 2000, in Northern Tuscany. Landslides, 3, 13-21

Cividini, A., Jurina L., Gioda, G., 1981, Some aspects of characterization problems in Geomechanics, International Journal of Rock Mechanics and Mining Sciences & Geomechanical Abstracts, 18(6), pp. 487-503 (1981), Elsevier.

Comegna, L., 2005, Proprietà e comportamento delle colate in argilla, tesi per il conseguimento del titolo di dottore di ricerca, Seconda Università degli Studi di Napoli, Italy 2005

Comegna, L., Picarelli, L., Urciuoli, G., 2007, The mechanics of mudslides as a cyclic undrained–drained process, Landslides Vol. 4, n. 3, 217-232

Conte, E., Dente, G., Guerricchio, A.,1991, Landslide movements in complex geological formations at Verbicaro, Southern Italy, Landslides, Bell (ed.), 1991, Balkema, Rotterdam, pp. 47-52

Conti, St. and Tosatti, G., 1994, Caratteristiche geologico-strutturali della Pietra di Bismantova e fenomeni franosi connessi (Appennino reggiano), Quaderni di geologia applicata , n. 1/1994

Corominas, J. and Ledesma, A., 2002, The role of geomorphological input in modelling of geomorphic processes. In: D. Delahaye, F. Levoy and O. Maquaire (Editors), Geomorphology: from expert opinion to modelling, Strasbourg, France

Corsini, A., 2000, L'influenza dei fenomini franosi sull'evoluzione geomorfologica post-glaciale dell'Alta Val Badia e della Valparola (Dolomiti), PhD thesis (2000), University of Bologna, Italy

Corsini, A., 2009 U, Field and monitoring data gathered at the Corvara landslide, Dolomites, Italy, unpublished manuscript

Corsini, A., Pasuto, A., Soldati, M., 1999, Geomorphological investigation and management of the Corvara landslide (Dolomites, Italy). Transactions-Japanese Geomorphological Union 20, 169–186

Corsini, A., Pasuto, A., Soldati, M., 2000, Landslides and Climate Change in the Alps Since the Late-Glacial: Evidence of Case Studies in the Dolomites (Italy), in: Landslides in research, theory and practice, Vol 1, edited by E. Bromhead, N. Dixon and M.-L. Ibsen, Proceedings of the 8th International Symposium on Landslides held in Cardiff on 26-30 June 2000, pp. 329-334

Corsini, A., Marchetti, M., Soldati, M., 2001, Holocene slope dynamics in the area of Corvara in Badia (Dolomites, Italy). Geografia Fisica e Dinamica Quaternaria 24, 127– 139

Corsini, A., Pasuto, A., Soldati, M., Zannoni, A., 2005, Field monitoring of the Corvara landslide (Dolomites, Italy) and its relevance for hazard assessment, Geomorphology Vol. 66, pp. 149-165 (2005), Elsevier

Corsini, A., Borgatti, L., Caputo, G., De Simone, N., Sartini, G., Truffelli, G., 2006, Investigation and monitoring in support of the structural mitigation of large slow moving landslides: an example from Ca' Lita (Northern Apennines, Reggio Emilia, Italy), Natural Hazards and Earth System Sciences, 6, 55–61, 2006, European Geosciences Union

Corsini, A., Borgatti, L., Cervi, F., Dahne, A., Ronchetti, F., Sterzai, P., 2009, Estimating mass-wasting processes in active earth slides – earth flows with time-series of High-Resolution DEMs from photogrammetry and airborne LiDAR. Natural Hazards and Earth System Science, 9, 2009, 2 ,433-439

Crosta, G. B., Imposimato, S., Roddeman, D. G., 2003, Numerical modelling of large landslides stability and runout. Natural Hazards and Earth System Sciences (2003) 3: 523–538

Crosta, G. B., Chen, H., Frattini, P., 2006, Forecasting hazard scenarios and implications for the evaluation of countermeasure efficiency for large debris avalanches, Engineering Geology, Volume 83, Issues 1-3, 28 February 2006, Pages 236-253, Elsevier

Crozier, M. J. and Glade, T., 2005, Landslide hazard and risk: Issues, Concepts and Approach, in: Glade, T., Anderson, M., Crozier M. (Eds.), - Landslide hazard and risk, Wiley, Chichester, 1-40

Cruden, D. M. and Varnes, D. J., 1996, Landslide types and processes, in: Turner A. K., Schuster R. L. (editors), Landslides: investigation and mitigation, Transportation Research Board, Special Report 247, National Academy Press, Washington D.C.

Cui, L. and Sheng, D., 2006, Genetic algorithms in probabilistic finite element analysis of geotechnical problems, Computers and Geotechnics Vol. 32, pp. 555-563 (2006).

Czurda, K. and Schütz, B. 2001, Die Großhangbewegung Sibratsgfäll-Rindberg (Vorarlberger Alpen). Ursachen, Prozesse und Langzeitverhalten. 13. Nat. Tagung Ingenieurgeol., Karlsruhe: 3 – 9. Essen: Glückauf

Dall`Olio, L., Ghirotti, M., Semenza, E., Turrini, M. C., 1988, The Tessina landslide (eastern Pre-Alps, Italy). Evolution and possible intervention methods. Proc. 5th Int. Symp. Landslides, Lausanne 2: 1317 – 1322. Rotterdam: Balkema

D'Elia, B. and Lanzo, G., Rossi-Doria, M., 2000, Slow Movements of Earthflow Accumulations along the Ionic Coast (South-Eastern Italy), in: Landslides in research, theory and practice, Thomas Telford, London, 2000

Dowding, C. H., Cole, R. G., Pierce, C. E., 2001, Detection of shearing in soft soils with compliantly grouted TDR cables. Proc. TDR 2001—Second International Symposium and Workshop on Time Domain Reflectometry for Innovative Geotechnical Applications, http://www.iti.Northwestern.edu/tdr/ tdr2001/proceedings/.

Duan, Q. Y., Gupta, V. K., Sorooshian, S., 1993, Shuffled Complex Evolution Approach for Effective and Efficient Global Minimization. Journal of Optimization Theory and Applications, Vol. 76, No. 3. S. 501 - 521

D.U.T.I., 1983, Projet d' école. Détection et utilisation des terrains instables. Rapport d'activité. Ecole polytechnique fédérale de Lausanne

Eberhardt, E., Stead, D., Coggan, J. S., 2004, Numerical analysis of initiation and progressive failure in natural rock slopes--the 1991 Randa rockslide, International Journal of Rock Mechanics and Mining Sciences, Volume 41, Issue 1, January 2004, Pages 69-87

Eberhart, R. C. and Kennedy, J., 1995, A new optimizer using particle swarm theory, Proceedings of the Sixth International Symposium on Micromachine and Human Science, pp. 30 - 43 (1995), Nagoya, Japan

Feng, X.-T., Chen, B.-R., Yang, C., Zhou, H., Ding, X., 2006, Identification of visco-elastic models for rocks using genetic programming coupled with the modified particle swarm optimization algorithm, International Journal of Rock Mechanics & Mining Sciences Vol. 43, pp. 789-801 (2006), Elsevier

Fetter, C. W., 1994, Applied hydrogeology, Published by Macmillan, New York

FGSV 1992, Merkblatt zur Felsbeschreibung für den Straßenbau, Forschungsgesellschaft für Straßen- und Verkehrswesen, Arbeitsgruppe Erd- und Grundbau, Köln, Ausgabe 1992, S. 27-37.

Finsterle, S., 1998, Multiphase Inverse Modelling: An Overview. U.S. Department of Energy's Geothermal Program Review XVI, Berkeley, California

Finsterle, S., 2000, Demonstration of Optimization Techniques for Groundwater Plume Remediation. Earth Sciences Division, Lawrence Berkeley National Laboratory, University of California, Berkeley

Finsterle, S., 2006, Demonstration of Optimization Techniques for Groundwater Plume Remediation using iTOUGH2, Environmental modelling and software, Vol. 22, pp 665-680 (2006), Elsevier

Firat, S., 2009, Stability analysis of pile-slope system, Scientific Research and Essay Vol. 4 (9), pp. 842-852, September, 2009, Academic Journals

Fischer, K., 1967, Erdströme in den Alpen. Mitt. Geogr. Gesellschaft München. 52: 231 – 246, München

Fleming, R.. W., Johnson, R. B., Schuster, R. L., 1988 A, The reactivation of the Manti landslide, Utah, in: The Manti, Utah, Landslide, US Geological Survey Professional Paper 1311-A

Fleming, R.. W., Schuster, R. L., Johnson, R. B., 1988 B, Physical properties and mode of failure of the Manti landslide, Utah, in: The Manti, Utah, Landslide, US Geological Survey Professional Paper 1311-B

Froldi, P., and Lunardi, P., 1994, Geometric and dynamic prperties of landslides in scaly clays in Northern Italy, 7th International IAEG Congress, 1994 Balkema, Rotterdam

Gattinoni, P., Scesi, L., Francani, V., 2005, Tensore di permeabilità e direzione di flusso preferenziale in un ammasso roccioso fratturato. Quaderni di Geologia Applicata 12-1, Pitagora Editrice, Bologna

Gens, A., Ledesma, A., Alonso, E. E., 1996, Estimation of parameters in geotechnical backanalysis - II. Application to a tunnel excavation problem, Computers and Geotechnics, Vol. 18, pp. 29-46 (1996), Elsevier

Ghirotti, M., Genevois, R., Teza, G., 2007, An example of a complex rock slope failure investigated by means of Laser Scanner Technique and numerical modelling, in: Geophysical Reasearch Abstracts, Vienna, European Geosciences Union, 2007, pp. 03957

Giusti, G., Iaccarino, G., Pellegrino, A., Russo, C., Urciuoli, G., Picarelli, L., 1996, Kinematic features of earth-flows in Southern Appenines, Italy. Proc 7th Int. Symp. Landslides, Tondheim 2: 457 – 462. Rotterdam: Balkema

Guadagno, F. M., Mele, R., Bassi, M. T., 1994, Slope instabilities and geological conditions in the Sant' Arcangelo Basin (Southern Italy), 7th International IAEG Congress 1994 Balkema, Rotterdam

Guerra, M., 2003, Studio della frana dei Boschi di Valoria (Comune di Frassinoro, Provincia di Modena), Tesi di laurea, Università degli Studi di Modena e Reggio Emilia

Haas, F., 2005 U, Katholische Universität Eichstätt, Lehrstuhl für Physische Geographie, Excel-file with precipitation data, meteorological station of the SEDAG project at Roter Graben, Lahnenwiesgraben valley, Landkreis Garmisch-Partenkirchen, Germany

Haefeli, R., 1967, Kriechen und progressiver Bruch in Schnee, Boden, Fels und Eis, in: Schweizerische Bauzeitung, 85. Jahrgang Heft 1, hrsg. v. d. Verlags-Aktiengesellschaft der Akademischen Technischen Vereine Zürich

Hanisch, J., 2004, Bedrohung des Lebensraums in den Hochgebirgen Asiens durch die Klimaerwärmung, Bodenseetagung 2004, Klimawandel: Thema für die Ingenieurgeologie?, Tagungsband, Bregenz 2004

Harp, E. L., Wells, G. W., Sarmiento, J. G., 1990, Pore pressure response during failure in soils. Geological Society of America Bulletin vol. 102-1, p. 428-438

Hazarika, H., Ozawa, T., Suzuki, Y., Okuzono, S., 2008, Three-Dimensional Numerical Simulations of Landslide for Slopes with Skewed Anchoring, Proceedings of the 12th Conference of the International Association for Computer Methods and Advances in Geomechanics IACMAG, 1-6 October, 2008, Goa, India

Hermanns, R. L., Blikra, L. H., Naumann, M., Nilsen, B., Panthi, K. K., Stromeyer, D., Longva, O., 2006, Examples of multiple rock-slope collapses from Kofels (Otz valley, Austria) and western Norway, Engineering Geology, Volume 83, Issues 1-3, Large Landslides: dating, triggering, modelling, and hazard assessment, 28 February 2006, Pages 94-108

Herzog, T., 1992, Geologische und ingenieurgeologische Untersuchungen der Gebiete Neuhüttenalm/Fockenstein und Ermoosgraben/Schliersee, Diplomarbeit am Institut für Geologie und Mineralogie Friedrich-Alexander-Universität Erlangen-Nürnberg

Hoek, E. and Brown, E. T., 1980, Underground excavations in Rock, Institution of Mining and Metallurgy, London.

Hoek, E. and Brown, E. T., 1997, Practical estimates of rock mass strength, International Journal of Rock Mechanics and Mining Sciences, Vol. 34, No. 8, pages 1165-1186, Elsevier

Hoek, E. and Diederichs, M. S., 2006, Empirical estimation of rock mass modulus, International Journal of Rock Mechanics and Mining Sciences 43: 203-215, Elsevier

Hofmann, R., 2009, Mechanik von Erd- und Schuttströmen, in: Marschallinger, R., Wanker, W., Zobl, F. (Hrsg.), Online Datenerfassung, berührungslose Messverfahren, 3D-Modellierung und geotechnische Analyse in Geologie und Geotechnik. Beiträge zur COG-Fachtagung Salzburg 2009, S. 202-210

Holtz, R. D. and Schuster, R. L., 1996, Stabilization of soil slopes. In: A.K. Turner and R.L. Schuster, eds. Landslides: investigation and mitigation. Special Report 247, 3, Washington, DC: National Academy Press, Transportation Research Board, 439-473

Hungr, O., Evans, S. G., Bovis, M. J., Hutchinson, J. N., 2001, A review of the classification of Landslides of the Flow Type, Environmental & Engineering Geoscience, v. 7, n. 3

Hutchinson, J. N., 1988, General report: Morphological and geotechnical parameters of landslides in relation to geology and hydrology. Proc. 5th Int. Symp. Landslides, Lausanne, 1: 3 – 35. Rotterdam: Balkema

Hutchinson, J. N. and Bhandari, R. K., 1971, Undrained Loading, a fundamental mechanism of mudflows and other mass movements, Géotechnique 21, No 4, 353-358

Iverson, R.. M.., 2005, Regulation of landslide motion by dilatancy and pore pressure feedback. Journal of Geophysical Research, 110, F2

Iverson, R.. M.., and Major, J. J., 1987, Rainfall, ground-water flow, and seasonal movement at Minor Creek landslide, northwestern California: Physical interpretation of empirical relations. Geological Society of America Bulletin. vol.99, pp. 579-594

Janbu, N., 1967, Settlement calculations based on the tangent modulus concept. Three guest lectures at Moscow State University. Bulletin No. 2 of Soil Mechanics and Foundation Engineering at the Technical University of Norway, Trondheim.

Jedlitschka, M., 1984, Untersuchung der Nährgebiete von Erdströmen in Hinblick auf deren Stabilisierung am Beispiel des Gschliefgrabens bei Gmunden, oberösterreich, Internationales Symposoium INTERPRAEVENT 1984, Villach, tagungspublikation, Bd. 2, s. 89-108

Kalkani, E. C., and Piteau, D. R., 1976, Finite element analysis of toppling failure at Hell's Gate Bluffs, British Columbia, Bull. Int. Assoc. of Eng. Geol. XIII (1976) (4), pp. 315–327

Keefer, D. K. and Johnson, A. M., 1983, Earth Flows: Morphology, Mobilization, and Movement, Geological Survey Professional Paper 1264, United States Government Printing office, Washington 1983

Keller, D., 2009, Analyse und Modellierung gravitativer Massenbewegungen in alpinen Sedimentkaskaden unter besonderer Berücksichtigung von Kriech- und Gleitbewegungen im Lockergestein (Lahnenwiesgraben, Garmisch-Partenkirchen), Doktorarbeit an der naturwissenschaftlichen Fakultät der Friedrich-Alexander-Universität Erlangen-Nürnberg

Keller, D., Heckmann, T., Moser, M., 2005, Felduntersuchungen und GIS-Analysen zu flachgründigen alpinen Schuttströmen bei Garmisch-Partenkirchen, in: Moser, M., 15. Tagung für Ingenieurgeologie, Erlangen, S. 189-194.

Kennedy, J. and Eberhart, R. C., 1995, Particle swarm optimization. Proceedings of the IEEE International Conference on Neural Networks, Piscataway, NJ: IEEE Press, 1942 –1948 (1995)

Kézdi, Á., 1968, Handbuch der Bodenmechanik, Band I, VEB Verlag für Bauwesen Berlin, Verlag der Ungarischen Akademie der Wissenschaften, Budapest 1968.

Klengel, K.. J., and Wagenbreth, O., 1982, Ingenieurgeologie für Bauingenieure, VEB Verlag für Bauwesen, Neuauflage Bauverlag GmbH Wiesbaden & Berlin.

Knabe, T., Schädler, W., Corsini, A., Mair, V., Schanz, T., 2009, Geotechnische Bewertung der Hangbewegung Corvara (Dolomiten, Italien) - Effizienz möglicher Sanierungsmaßnahmen -Beiträge zum 24. Christian Veder Kolloquium, 16. und 17. April 2009, "Stabilisierung von Rutschhängen, Böschungen und Einschnitten" - Heft 35, 2009, GGG Mitteilungshefte, Gruppe Geotechnik Graz, Institut für Bodenmechanik und Grundbau, Technische Universität Graz, Austria.

Krahn, J., 2004, Seepage Modeling with SEEP/W. GEO-SLOPE International Ltd.

Krahn, J., and Morgenstern, N. R., 1976, The mechanics of the Frank Slide, Rock Engineering for Foundations and Slopes, Proc. of ASCE Geotech. Eng. Specialty Conf., Boulder, ASCE, New York (1976), pp. 309–332

Ledesma, A., Gens, A., Alonso, E. E., 1996 A, Estimation of Parameters in Geotechnical Backanalysis– I. Maximum Likelihood Approach, Computers and Geotechnics, Vol. 18, No.1, pp. 1-27 (1996), Elsevier.

Ledesma, A., Gens, A., Alonso, E. E., 1996 B, Parameter and Variance Estimation in Geotechnical Backanalysis using Prior Information, International Journal for Numerical and Analytical Methods in Geomechanics, Vol. 20, pp. 119-141 (1996)

Lee, C. H. and Farmer, I., 1993, Fluid flow in discontinuous rocks, Published by Chapman and Hall, London

Le Maitre, R. W., 2002, Igneous rocks, IUGS classification and glossary, recommendations of the International Union of Geological Sciences, Subcommission on the systematics of Igneous Rocks, 2. Aufl., New York, cambridge University Press 2002

Schädler, W., Lemke, K., Lorenz, H., Klimitsch, A., Köditz, J., Schäfer T., Witt, K. J., 2006 U, Results of laboratory study on earth-flow-materials, unpublished Technical Report, Material Research and Testing Institute MFPA, Weimar, Germany, 2006

Leroueil, S., Locat, J., Vaunat, J., Picarelli, L., Lee, H., Faure, R.., 1996, Geotechnical characterization of slope movements, Landslides, Senneset (ed.) 1996 Balkema, Rotterdam, pp. 53-74

Levasseur, S., Malecot, Y., Boulon, M., Flavigny, E., 2007 A, Soil parameter identification using a genetic algorithm, International Journal for Numerical and Analytical Methods in Geomechanics (2007), John Wiley & Sons

Levasseur, S., Malecot, Y., Boulon, M., Flavigny, E., 2007 B, Soil parameter identification from in situ measurements using a genetic algorithm and a principle component analysis, Proceedings of the 10th NUMOG conference, Rhodes (2007)

Mackey, B. H., Roering, J. J., McKean, J. A., 2009, Long-term kinematics and sediment flux of an active earthflow, Eel River, California. Geology, September 2009, v.37, no.9, p.803-806, Geological Society of America

Malecot, Y., Flavigny, E., Boulon, M., 2004, Inverse Analysis of Soil Parameters for Finite Element Simulation of Geotechnical Structures:Pressuremeter Test and Excavation Problem, in: Brinkgreve, Schad, Schweiger & Willand (eds.) Proc. Symp. Geotechnical Innovations, pp. 659-675. (2004), Verlag Glückauf, Essen

Malet, J.-P., Laigle, D., Remaitre, A., Maquaire, O., 2005, Triggering conditions and mobility of debris flows associated to complex earthflows. Geomorphology, Volume 66, Issues 1-4

Manly, B. F. J., 1944, Multivariate statistical methods - a primer, 3rd Edition (1944), Chapman & Hall / CRC

Manzi, V., Leuratti, E., Lucente, C. C., Medda, E., Guerra, M., Corsini, A., 2004, Historical and recent hydrogeological instability in the Monte Modino area: Valoria, Tolara and Lezza Nuova landslide reactivations (Dolo-Dragone valleys, Modena Apennines, Italy), GeoActa 3:1-13

Marcato, G., Mantovani, M., Fujisawa, K., Pasuto, A., Tagliavini, F., Silvano, S., Zabuski, L., 2008, Monitoring, modelling and mitigation of the Moscardo landslide (Eastern Italian Alps), Geophysical Research Abstracts, Vol. 10, EGU 2008-A-04811, EGU General Assembly, European Geosciences Union, Vienna 2008

Marcato, G., Mantovani, M., Pasuto, A., Tagliavini, F., Silvano, S., Zabuski, L., 2009, Assessing the possible future development of the Tessina landslide using numerical modelling, Landslide Processes: from geomorphological mapping to dynamic modelling, Strasbourg

Marinos, P. and Hoek, E., 2001, Estimating the geotechnical properties of heterogeneous rock masses such as flysch, Bulletin of Engineering Geology and the Environment, 60: 85-92

Meier, H., 2005 U, Pore pressure data from earthflow nr. 13 in the Lahnenwiesgraben valley, Landkreis Garmisch-Partenkirchen, Bavarian Alps, Excel Files, unpublished

Meier, J., 2008, Parameterbestimmung mittels inverser Verfahren für geotechnische Problemstellungen, Dissertation an der Fakultät Bauingenieurwesen, Bauhaus-Universität Weimar, Weimar, Germany

Meier, J., 2009 U, Statistical data and optimisation results for simulations of laboratory tests and slope movements, case study of Corvara, Dolomites, Italy, Excel-, emf- and xml-files, unpublished

Meier, J., Rudolph, S., Schanz, T., 2006, Effektiver Algorithmus zur Lösung von inversen Aufgabenstellungen - Anwendung in der Geomechanik, Bautechnik, 83, Heft 7 (2006), Ernst & Sohn, Berlin.

Meier, J., Schädler, W.; Borgatti, L.; Corsini, A.; Schanz. T., 2008, Inverse Parameter Identification technique using PSO Algorithm applied to Geotechnical Modeling. Journal of Artificial Evolution and Applications, http://www.hindawi.com.

Mesri, G. and Godlewski, P., 1977, Time- and Stress-Compressibility Interrelationship; Journal of the geotechnical engeniering division 103, 417-430, NRC Canada

Mesri, G. and Castro, A., 1987, C_α/C_C Concept and K_0 during Secondary Compression; Journal of geotechnical engineering 112(3), 230 – 247

Mitchell, J. K. and Soga, K. 2005, Fundamentals of soil behavior, third edition 2005, John Wiley & Sons, inc. Hoboken, New Jersey

Moser, M., 2002, Geotechnical aspects of landslides in the Alps, Landslides, Rybár, Stemberk & Wagner (eds) 2002 Swets & Zeitlinger, Lisse, pp. 23-43

Moser, M. and Üblagger, G., 1984, Vorschläge zur Erstellung geotechnischer Karten und Erhebungen im Rahmen von Gefahrenzonenfällen in Hangbereichen, Int. Symp. Interpraevent 1984, 2, S. 275-287, Villach

Nozaki, T., Miyazava, M., Uchida, M., Okada, Y., 1994, Earthflow induced by snow load in Nagano Prefecture, Central Japan. 7th International IAEG Congress. Balkema, Rotterdam, p. 1431-1439

Panizza, M., Pasuto, A., Silvano, S., Soldati, M., 1996, Temporal occurrence and activity of landslides in the area of Cortina d'Ampezzo (Dolomites, Italy), Geomorphology 15 (1996) 311-326

Panizza, M., Pasuto, A., Silvano, S., Soldati, M., 1997, Landsliding during the Holocene in the Cortina d'Ampezzo region, Italian Dolomites, Paläoklimaforschung – Palaeoclimate Research, 19, 17-31, Gustav-Fischer-Verlag

Panizza, M., Silvano, S., Corsini, A., Soldati, M., Marchetti, M., Borgatti, L., Ghinoi, A., Piacentini, D., Pasuto, A., Zannoni, A., Marcato, G., Mantovani, M., Tagliavini, F., Moretto, S., 2006, Definizione della pericolosità e di possibili interventi di mitigazione della frana di Corvara in Badia. Provincia Autonoma di Bolzano - Alto Adige; Open File Report Autonome Provinz Bozen - Südtirol

Pellegrino, A., Ramondini, M., Russo, C., Urciuoli, G., 2000, Kinematic Features of Earthflows in Southern Italy, Landslides in research, theory and practice, Thomas Telford, London, 2000

Pentecost, A., 2005, Travertine, Dordrecht, Netherlands, Kluwer Academic Publishers Group

Pettijohn, F. J. 1975, Sedimentary Rocks, Harper & Row

Picarelli, L., Russo, C., Urciuoli, G., 1995, Modelling earthflow movement based on experiences, in: Danish Geotechnical Society (ed) Proceedings of the 11th European conference on soil mechanics and foundation engineering, Copenhagen, Vol 6, pp. 157-162

Picarelli, L., Oboni, F. S. Evans, S. G., Mostyn, G., Fell, R.. 2005, State of the Art Paper 2, Hazard characterization and quantification. Int. Workshop on Landslide Risk Mitigation, Vancouver: 27-61

PLAXIS V8, Reference Manual, 2003, edited by R.B.J. Brinkgreve, Technical University of Delft and PLAXIS BV, The Netherlands, in cooperation with R. Al-Khoury, K.J.Bakker, P.G. Bonnier, P.J.W. Brand, W.Broere, H.J.Burd, G.Soltys, P.A.Vermeer, C. Vogt-Breier, D. Waterman, DOC Den Haag, PLAXIS BV, Delft

PLAXIS Version 8 Material Models Manual, downloaded at the official website of PLAXIS BV: http://www.plaxis.nl, download date: 21.01.2005.

Poeter, E. P. and Hill, M. C., 1997, Inverse models: a necessary next step in ground-water modeling. Ground Water, 35:250-260.

Poschinger, A. von, 1992, Die jüngsten Hangbewegungen bei Hutterer, Gemeinde Inzell, und ihre Bedeutung für die Aufnahme von Hangbewegungen im Bayerischen Alpenraum. Z. angew. Geol. 38: 21 – 25

Radbruch-Hall, D. H., Varnes D. J., Savage, W. Z., 1976, Gravitational spreading of steep sided ridges (sackung) in western United States, Int. Assoc. Eng. Geol., Bull. 14 (1976), pp. 23–35

Raetzo, H. and Lateltin, O., 1996, Rutschung Falli Hölli, ein ausserordentliches Ereignis?, Internationales symposion INTERPRAEVENT 1996 – Garmisch-Partenkirchen, Tagungspublikation, Bd. 3, S. 129-140

Raetzo, H., Keusen, H. R., Oswald, D., 2000, Rutschgebiet Hohberg-Rohr (Plaffeien, FR) – Disposition und Aktivität, Bull. Angew. Geol., Vol. 5, Nr. 1, S. 55-68

Ressle, A., 2006, Diplomarbeit am Institut für Geologie und Mineralogie Friedrich-Alexander-Universität Erlangen-Nürnberg

Rohn, J., 1991, Geotechnische Untersuchungen an einer Großhangbewegung in Bad Goisern (Oberösterreich). Schr. Angew. Geol. Karlsruhe 14: XVIII, 193 p. Karlsruhe

Rohn, J., Resch, M., Schneider, H., Fernandez-Steeger, T. M., Czurda, K., 2004, Large-scale lateral spreading and related mass movements in the Northern Calcareous Alps, Bull. Eng. Geol. Env. (2004) 63:71-75, Springer-Verlag

Rohn, J., Ehret, D., Moser, M., Czurda, K., 2005, Prehistoric and recent mass movements of the World Cultural Heritage Site Hallstatt, Austria, Environ Geol (2005) 47: 702-714

Ronchetti, F., 2007, Caratteristiche idro-meccaniche di grandi frane per scivolamento-colata in ammassi rocciosi deboli ed eterogenei: analisi e modellizzazione di casi di studio circostanti il Monte Modino (Alta Val Secchia, Appennino settentrionale), Tesi di Dottorato in Scienze della Terra, Dipartimento di Scienze della Terra, Università degli Studi di Modena e Reggio Emilia

Ronchetti, F., Borgatti, L., Cervi, F., Corsini, A., Pellegrini, M., Ambrosini, L., Campagnoli, I., Gatti, A., Lorenzi, L., Mazzini, A., 2006, Studio e monitoraggio idrogeologico delle frane "Lezza Nuova" e "Tolara" in comune di Frassinoro e Montefiorino (Provincia di Modena), Relazione Tecnica Finale, Dipartimento di Scienze della Terra, Università di Modena e Reggio Emilia

Ronchetti, F., 2007 U, Steady-State Groundwater Flow Calculation and Infiltration and Seepage Model for source area S3 of the Corvara earthflow (Dolomites, Italy), applying the Finite-Element Code SEEP/W, unpublished manuscript

Ronchetti, F, 2008 U, Geomorphological evolution and reactivation events of the Valoria earthflow, upper Secchia valley, Northern Apennines. Unpublished manuscript

Ronchetti, F, 2009 U, Laboratory data and monitoring data on movements and hydrogeology of the Valoria earthflow, upper Secchia valley, Northern Apennines. Unpublished manuscript

Ronchetti, F., Borgatti, L., Cervi, F., Lucente, C. C., Veneziano, M., Corsini, A., 2007, The Valoria landslide reactivation in 2005-2006 (Northern Apennines, Italy), Landslides, Volume 4, Number 2, pages 189-195, May 2007, Recent Landslides, Springer Verlag Berlin/Heidelberg

Ronchetti, F., Borgatti, L., Cervi, F., Corsini, A., 2009 A, Hydro-mechanical features of landslide reactivation in weak clayey rock masses. Bulletin of Engineering Geology and the Environment, DOI 10.1007/s10064-009-0249-3.

Ronchetti, F., Borgatti, L., Cervi, F., Gorgoni, C., Piccinini, L., Vincenzi, V., Corsini, A., 2009 B, Groundwater processes in a complex landslide, northern Apennines, Italy. Natural Hazards and Earth System Sciences, 9, 895-904

Samper-Calvete, F. J. and Garcia-Vera, M. A., 1998, Inverse modeling of groundwater flow in the semiarid evaporitic closed basin of Los Monegros, Spain. Hydrogeology Journal, 6:33-49.

Sartini, G., Caputo, G., De Simone, N., Truffelli, G., Borgatti, L., Cervi, F., Corsini, A., Ronchetti, F., 2007, Consolidamento di versanti instabili mediante opere di contenimento strutturale e tecniche di drenaggio profondo: gli esempi di Acquabona, Cervarezza, Magliatica e Ca' Lita (Appennino reggiano), Giornale di Geologia Applicata 7 (2007) 5-16.

Savage, W. Z., Varnes, D. J., Schuster, R. L., Highland, L. M., 1996, The Slumgullion earth flow, southwestern Colorado, landslides, senneset (ed.) 1996 Balkema, Rotterdam

Schanz, T., Zimmerer, M. M., Datcheva, M., Meier, J., 2006, Identification of Constitutive Parameters for Numerical Models via Inverse Approach, Felsbau, Vol. 24 No. 2, pp. 11-21 (2006)

Scheidegger, A. E., 1975, Physical Aspects of Natural Catastrophes, Elsevier, Amsterdam

Schulz, W. H., McKenna, J. P., Kibler, J. D., Biavati, G., 2009, Relations between hydrology and velocity of a continuously moving landslide—evidence of pore-pressure feedback regulating landslide motion? Landslides 6:181–190

Simoni, A. and Berti, M., 2007, Transient hydrological response of weathered clay shale and its implications for slope instability. Proceedings of the First North American Landslide Conference, Vail, Colorado, June 3-8, 2007

Simoni, A., Berti, M., Generali, M., Elmi, C., Ghirotti, M., 2004, Preliminary result from pore pressure monitoring on an unstable clay slope. Engineering Geology 73, 117– 128

Soldati, M., Corsini, A., Pasuto, A., 2004, Landslides and climate change in the Italian Dolomites since the Late Glacial, Catena, Vol. 55, pp. 141-161 (2004), Elsevier

Soldati, M., Borgatti, L., Cavallin, A., De Amicis, M., Frigerio, S., Giardino, M., Mortara, G., Pellegrini, G. B., Ravazzi, C., Surian, N., Tellini, C., con la collaborazione di Alberto, W., Albanese, D., Chelli, A., Corsini, A., Marchetti, M., Palomba, M., Panizza, M., 2006, Geomorphological evolution of slopes and climate changes in Northern Italy during the late Quaternary: spatial and temporal distribution of landslides and landscape sensitivity implications, Geografia Fisica e Dinamica Quaternaria 29(2), pp. 165-183

Spickermann, A., 2008, Analyse tiefgreifender Hangdeformationen – Einfluss des Initialspannungszustands und der konstitutiven Formulierung, Dissertation an der Fakultät Bauingenieurwesen, Bauhaus-Universität Weimar

Spickermann, A., Schanz, T., Datcheva, M, 2003, Analysis of a deep seated slope failure in the Alps. AK 1.6 Workshop: Nachweise für Böschungen und Baugruben mit numerischen Methoden, Bauhaus-Universität Weimar, Schriftenreihe Geotechnik, Heft 11, pp. 47-58, 2003

Stead, D., Eberhardt, E., Coggan, J. S., 2006, Developments in the characterization of complex rock slope deformation and failure using numerical modelling techniques. Engineering Geology, 83, 217-235

Stuiver, M., Reimer, P. J., Bard, E., Beck, J. W., Burr, G. S., Hughen, K. A., Kromer, B., MacCormac, G., Van Der Plicht, J., Spurk, M., 1998, INTCAL98 Radiocarbon age calibration 24,000-0 cal BP, Radiocarbon Vol. 40, No. 3, pp 1041-1083

Tarchi, D., Casagli, N., Fanti, R., Leva, D. D., Luzi, G., Pasuto, A., Pieraccini, M., Silvano, S., 2003, Landslide monitoring by using ground-based SAR interferometry: an example of application to the Tessina landslide in Italy, Enineering Geology 68 (2003) 15-30, Elsevier

UNESCO Working Party on World Landslide Inventory, 1993, A suggested method for describing the activity of a landslide, Bulletin of the International Association of Engineering Geology, No. 47, Paris

van Asch, T. W. J., Malet, J. P., van Beek, L. P. H., Amitrano, D., 2007 A, Techniques, issues and advances in numerical modelling of landslide hazard. Bulletin de la Societe Geologique de France, 178, 65-88

van Asch Th. W. J., van Beek L. P. H., Bogaard T. A., 2007 B, Problems in predicting the mobility of slow-moving landslides. Eng Geol 91:46–55

van Beek, L. P. H., 2002, Assessment of the influence of changes in climate and land use on landslide activity in a Mediterranean environment. PhD thesis, Neth. Geogr. Stud. 294, Utrecht, KNAG / Faculty of Geographical Sciences

van Beek, L. P. H., and Van Asch, Th. W. J., 2004, Regional Assessment of the Effects of Land-Use Change on Landslide Hazard by Means of Physically Based Modelling, Natural Hazards 31: 289–304, 2004, Kluwer Academic Publishers, The Netherlands.

van den Ham, G., 2006, Numerical simulation and engineering-geological assessment of a creeping slope in the Alps, Dissertaion an der Fakultät für Bauingenieur-, Geo- und Umweltwissenschaften der Universität Karlsruhe (TH) 2006

van den Ham, G., Rohn, J., Meier, T., Czurda, K., 2009, Finite Element simulation of a slow moving natural slope in the Upper-Austrian Alps using a visco-hypoplastic constitutive model, Geomorphology, Volume 103, Issue 1, Dating, triggering, modelling, and hazard assessment of large landslides, 1 January 2009, Pages 136-142

VanDine, D. F., 1983, Drynoch landslide, British Columbia – A history. Can. Geotechn. Journ. 1 (20): 82 – 103

van Westen, C. J., van Asch, Th. W. J., Soeters, R., 2006, Landslide hazard and risk zonation -why is it still so difficult? – Bull. Eng. Geol. Env. Vol. 65 nr. 2, May 2006, pp. 167-184

Varnes, D. J., 1978, Slope movement types and processes. In: R. L. Schuster & R. J. Krizek (Editors), Landslides: analysis and control, Transportation Research Board, National Research Council, Special report 176, Washington D.C.

226

Vermeer, P. A. and Neher, H. P., 1999, A soft soil model that accounts for creep, Proceedings of the International Symposium "Beyond 2000 in Computational Geotechnics – 10 Years of PLAXIS International", Balkema, Amsterdam

VERSINCLIM, 1998, Comportement passé, présent et futur des grands versants instables subactifs en fonction de l'évolution climatique, et évolution en continu des mouvements en profondeur, Rapport final du Programme National de Recherche "Changements climatiques et catastrophes naturelles" PNR 31, vdf Hochschulverlag AG an der ETH Zürich 1998

Vulliet, L. and Bonnard, Ch., 1996, The Chlöwena landslide: Prediction with a viscous model, Landslides, Senneset (ed.), 1996 Balkema, Rotterdam

Wasowski, J., 1998, Inclinometer and piezometer record of the 1995 reactivation of the Acquara-Vadoncello landslide, Italy, 8th International IAEG Congress, 1998 Balkema, Rotterdam, pp. 1697-1704

Weidinger, J. T., 2009, Das Gschliefgraben-Rutschgebiet am Traunsee-Ostufer (Gmunden/OÖ) – Ein Jahrtausende altes Spannungsfeld zwischen Mensch und Natur, in: Jahrbuch der Geologischen Bundesanstalt, Bd. 149, Heft 1, S. 195-206, Wien

Will, J., Roos, D., Riedel, J., Bucher, C., 2003, Robustness Analysis in Stochastic Structural Mechanics, NAFEMS Seminar - Use of Stochastics in FEM Analyses (2003), Wiesbaden

Williams, G. P., 1988, Stream-channel changes and pond formation at the 1974-76 Manti Landslide, Utah, in: The Manti, Utah, Landslide, U. S. Geological Survey professional Paper 1311-C

Zhang, Z. F., Ward A. L., Gee, G. W., 2003, Estimating Soil Hydraulic Parameters of a Field Drainage Experiment Using Inverse Techniques, Vadose Zone Journal, Vol. 2, pp. 201-211 (2003)

Zoebisch, M. A. and Johansson, A., 2002, Erosion scars caused by earth flows – a case study from central Kenya, 12th ISCO Conference, Beijing 2002